Practical Feline Behaviour
Understanding Cat Behaviour and Improving Welfare

The appendices for this book are available at
https://www.cabi.org/openresources/47838

Practical Feline Behaviour

Understanding Cat Behaviour and Improving Welfare

Trudi Atkinson

Clinical Animal Behaviourist, UK

CABI

CABI is a trading name of CAB International

CABI	CABI
Nosworthy Way	745 Atlantic Avenue
Wallingford	8th Floor
Oxfordshire OX10 8DE	Boston, MA 02111
UK	USA
Tel: +44 (0)1491 832111	Tel: +1 (617)682-9015
Fax: +44 (0)1491 833508	E-mail: cabi-nao@cabi.org
E-mail: info@cabi.org	
Website: www.cabi.org	

© T. Atkinson, 2018. All rights reserved. No part of this publication may be reproduced in any form or by any means, electronically, mechanically, by photocopying, recording or otherwise, without the prior permission of the copyright owners.

A catalogue record for this book is available from the British Library, London, UK.

Library of Congress Cataloging-in-Publication Data

Names: Atkinson, Trudi, author. | C.A.B. International, issuing body.
Title: Practical feline behaviour : understanding cat behaviour and improving welfare / Trudi Atkinson.
Description: Oxfordshire, UK ; Boston, MA : CABI, [2018] | Includes bibliographical references and index.
Identifiers: LCCN 2017060813 (print) | LCCN 2017061262 (ebook) | ISBN 9781780647821 (pdf) | ISBN 9781780647814 (ePub) | ISBN 9781780647838 (pbk. : alk. paper)
Subjects: | MESH: Cats | Behavior, Animal | Animal Welfare
Classification: LCC SF446.5 (ebook) | LCC SF446.5 (print) | NLM SF 446.5 | DDC 636.8--dc23
LC record available at https://lccn.loc.gov/2017060813

ISBN-13: 9781780647838 (pbk)
          9781780647821 (PDF)
          9781780647814 (ePub)

Commissioning editor: Caroline Makepeace
Editorial assistant: Alexandra Lainsbury
Production editor: Tim Kapp

Typeset by SPi, Pondicherry, India
Printed and bound in Great Britain by Severn, Gloucester.

Contents

About the Author	xi
Foreword	xiii
Preface	xv
Acknowledgements	xvii

PART ONE: UNDERSTANDING FELINE BEHAVIOUR

1. The Origin and Evolution of the Domestic Cat	3
The African Wildcat	3
Domestication	5
Historical Attitudes Towards Cats	7
The Modern-day Pet Cat	8
Current Attitudes Towards Cats and Keeping Cats as Pets	10
Feline Behaviour	11
2. The Senses	14
Sight	14
Hearing	16
Olfaction (Sense of Smell)	17
Touch	18
Balance	19
3. Feline Communication	22
Vocalizations	22
Visual Signalling	26
Tactile Communication	33
Olfactory Communication	35
4. Social, Feeding and Predatory Behaviour	42
Social Behaviour	42
Feeding Behaviour	51
Predatory Behaviour	52
5. Kitten to Cat – Reproduction and the Behavioural Development of Kittens	58
Reproductive Behaviour of Tomcats	58
Reproductive Behaviour of the Queen	60
Physical and Behavioural Development of Kittens	66
6. Health and Behaviour	79
Pain	79
Disease	81

Old Age	81
Stress	82
The Impact of Stress on Physical Health	84
The Impact of Stress on Mental and Emotional Health	90
Acute Stress – the Influence on Physiological Parameters	92
Assessing Stress	92
Observable Signs of Stress in Cats	93

7. Learning, Training and Behaviour — 98
- Why Train Cats? — 98
- Learning Theory — 98
- Habituation — 98
- Sensitization — 99
- Associative Learning — 100
- Factors Influencing Learning — 112

PART TWO: PRACTICAL FELINE BEHAVIOUR

8. Advice for Breeders — 119
- The Responsibility of the Breeder in the Prevention of Behaviour Problems — 119
- Selection of Queen and Stud — 119
- Pregnancy — 120
- Parturition — 123
- Pre-weaning Period — 123
- Weaning — 124
- Early Experience — 126
- Maternal Aggression — 129
- Education of New Owners — 129

9. Advice for Prospective Cat Owners — 131
- Is a Cat the Best Pet for You? — 131
- Making the Right Choice — 133
- The Importance of Early-life Influences — 136
- What to Look For and What to Avoid — 136
- Bringing your New Cat or Kitten Home — 138

10. Advice for Cat Owners — 141
- Managing Feline Stress — 141
- Indoor Cat or Outdoor Access? — 141
- A Fenced-off Area — 142
- Cat Flaps — 145
- Neutering — 145
- Other Cats — 146
- Cats, Babies and Children — 149
- Cats and Dogs — 149
- Avoiding House-training Issues — 149
- Preventing Human-directed Aggression — 150
- Keeping Your Cat Healthy — 151

11. Advice for Veterinary Professionals — 152
   a. Part 1 – The Cat in the Veterinary Clinic — 152
      The Journey to the Practice — 153
      The Waiting Room — 153
      Appointments — 154
      Greeting and Speaking to Clients — 154
      The Consultation Room — 155
      Handling and Examining — 155
      Hospitalization — 158

   b. Part 2 – Advising Clients: Prevention and Treatment of Feline Behaviour Problems — 164
      Talks — 164
      Nurse/Technician Behavioural Clinics — 164
      Handouts — 165
      Behavioural First Aid — 165
      Referral or Treat 'In-house' — 167
      Who to Refer to? — 168
      Behavioural Pharmacology — 169
      Pheromonatherapy — 170
      Complementary and Alternative Medicine (CAM) — 174
      The Placebo Effect — 178

12. Advice for Other Cat Carers — 180
   General Advice for All — 180
   Advice for Shelters and Catteries — 180
   Foster Care — 187
   Adoption — 188
   Cat Sitters (Advice for Sitters and Owners) — 189
   Cat Cafés — 190
   An Alternative Idea — 194

Appendices

1. Environmental Enrichment — 195
   Space — 195
   Hiding Areas — 195
   Food Foraging and Food Puzzles — 195

2. Play — 200
   Social Play — 200
   Object Play — 201
   How to Tell if the Cat is Interested in Play — 202
   Ending the Game — 204
   Catnip — 204

3. Reducing Resource Competition in a Multi-cat Household — 205
   Food — 205
   Water — 206
   Resting Places — 206
   Litter Trays — 206

4. Neutering	207
Spay	207
Castration	207
Why Neuter?	207
When to Neuter	208
5. Introducing an Additional Cat to your Household	210
Points to Consider Before Getting Another Cat	210
Correct Introductions: Increasing the Chances that a New Cat and Resident Cat(s) will Accept Each Other	211
Keeping the Peace	214
6. Helping a Grieving Cat	215
Do Cats Grieve?	215
How Can We Help?	215
7. Introducing Cats and Dogs	217
If You Have a Dog and are Considering Getting a Cat	217
If You Have a Cat and are Considering Getting a Dog	217
Introducing a New Cat or Kitten into a Home with a Resident Dog	218
Introducing a Dog or Puppy into a Home with a Resident Cat	221
Using a Crate	222
8. Cat Flaps	223
Where to Fit a Cat Flap	223
Types of Cat Flap	224
Training Your Cat to Use a Cat Flap	225
9. Teaching Your Cat to Come to You When You Call	226
10. Cats, Babies and Children	227
A New Baby	227
Cats and Children	228
11. House-training Your Cat or Kitten	230
Insufficient Number of Litter Trays	230
The Location of the Litter Trays	230
The Size and Shape of the Litter Tray	231
Covered vs Uncovered Litter Tray?	231
The 'Wrong' Type of Litter Substrate	231
Using an 'Outdoor' Toilet	232
House-training 'Accidents'	233
If the Problem Persists	233
House-Training Problem or Scent-Marking?	233
12. Training Your Cat to Like the Cat Carrier	235
13. Medicating Your Cat	240
Mixing in Food	240
Physically Medicating the Cat	241
Applying a Spot-on Treatment	242

14. Teaching your Cat or Kitten to Accept Veterinary Examination	243
Examining the Mouth	243
Examining the Ears	244
Examining the Feet and Clipping the Nails	244
15. First Aid Advice for Common Feline Behaviour Problems	245
General Advice for all Problems	245
House Soiling	245
Aggression to People	247
Fighting in a Multi-cat Household	249
Furniture Scratching	250
16. Friend or Foe	251
Signs of a Friendly Relationship	251
Signs of a Bad Relationship	251
Signs of a 'Tolerant' Relationship	252
17. Approaching, Stroking and Picking Up	253
The Approach	253
Picking the Cat Up	254
18. Recognizing Stress	256
Physiological Signs of Stress	256
Behavioural Signs of Stress	256
Glossary	259
List of Useful Websites	261
Recommended Reading List	263
Index	265

About the Author

Even as a small child Trudi wanted to work with animals and if asked what she wanted to be when she grew up her standard answer was 'a zoologist'. That was until she was informed that she would need to stay on at school and continue studying until she was at least 18! After that she gave up on lofty academic ambitions, although the desire to work with animals in one way or another continued. After leaving school (at 17) and after having a variety of jobs, including some time as a zookeeper, she entered veterinary practice in 1983 and qualified as a veterinary nurse in 1986.

She continued working in veterinary practice for 17 years in total, and it was while working as a nurse that her interest in companion animal behaviour developed. She undertook extra study and in 1999 she gained the diploma in advanced studies in Companion Animal Behaviour Counselling from the University of Southampton. That same year she was accepted as a full member of the Association of Pet Behaviour Counsellors (APBC) and in 2003 achieved certification as a Clinical Animal Behaviourist (CCAB) under the accreditation scheme run by the Association for the Study of Animal Behaviour (ASAB).

Although having previously worked with both dogs and cats, feline behaviour and welfare has always been her primary area of interest and is now her sole focus.

Trudi runs a feline behaviour referral practice covering the South West of England, and in 2018 she was recognized as a Certified Cat Consultant by the International Association of Animal Behavior Consultants (IAABC). She has also presented numerous talks and contributed written articles and book chapters for organizations including the British Small Animal Veterinary Association (BSAVA), British Veterinary Nursing Association (BVNA), British Veterinary Behaviour Association (BVBA) and the Association of Pet Behaviour Counsellors (APBC).

To keep up to date with the latest research, continued professional development is a necessary requirement for any clinical animal behaviourist and something that is particularly important in the continually evolving field of feline ethology. So at more than a few decades past 18 years of age Trudi is still studying!

Foreword

There are hundreds of books about dog behaviour, training and behavioural disorders: why are there so few about cats? Are cats simply not worthy of their owners' understanding, or are there other factors involved?

The discrepancy doesn't reflect their popularity: there are roughly equal numbers of pet cats and dogs in both the UK and the USA. Could it be that dogs have more behavioural disorders than cats do? That depends on precisely how a "disorder" is defined, but the little information that does exist suggests that about half the cats in the UK regularly behave in ways that either indicate that their welfare is compromised, or their owners find objectionable (or both – see Chapter 1). Thus the issues addressed in this book are far from esoteric, applying to tens of millions of cats, and it's arguable that every one of those cats would be better off if their owner were to obtain a copy and read it from cover to cover. So what is it about cats that persuades their owners that they need less help than dogs?

Partly, it must be down to the typical cat's personality – or at least the traditional interpretation of the way they go about their everyday lives. Cats are generally portrayed as independent animals, often characterised as 'aloof', 'solitary', even 'antisocial'. Historically, such attitudes have their origins in the domestic cat's traditional role as rodent exterminators. Even as recently as half a century ago, most pet cats were allowed to roam around and obtain some of their food by hunting, because that was regarded as part of their nature, something they could only fulfil if they were left to their own devices.

There is biological logic behind such preconceptions. Unlike dogs, which form such strong attachments to their owners that they will literally follow them anywhere, cats form their strongest attachments to the places in which they live. That's not to say that they are incapable of forming affectionate relationships with their owners, but a secure place to live always comes first – hence the recommendation to keep cats indoors for at least two weeks following a house move, which is evidently the amount of time it takes for the cat to forget its attachment to the old house and learn that the new one has everything it needs. Many of the behavioural disorders discussed in this book stem from the cat not feeling as secure in its surroundings as it would like to be.

Today, the expectations that owners have for their cats have changed. Many cats live in apartments, many are confined indoors for their entire lives. Hunting is discouraged as unnecessarily bloodthirsty, messy and cruel, not to mention its supposed effects on wildlife populations. The sexual exploits of cats, once sufficiently commonplace to provide metaphors for human behaviour, are now rarely witnessed, due to the widespread adoption of neutering as the approved method for regulation of their numbers: drowning of unwanted litters, unremarkable a century ago, could now earn a prison sentence for the perpetrator.

Meanwhile the cats themselves have hardly changed at all under the skin, and so need their owners' help to cope with the pressures of twenty-first century life. Even today, most pet cats are the product of unplanned matings between individuals that have somehow escaped being neutered, preventing any change in the underlying genetics that give rise to territorial and hunting behaviour. Thus the only way cats can adapt to the various demands that owners now place on them is through learning, and many of the issues described in this book stem from when they fail to adjust. Many of these are preventable, if only owners, breeders and veterinary professionals understood more about cats' psychology and how best their needs can be met.

There is a plethora of advice about dog behaviour – some of it, sadly, contradictory – but rather little for cat owners. This book is a most welcome addition to the field, one that every cat would – if it only knew how – thank its owner for reading.

John Bradshaw
March 2018

Preface

The old saying 'The more you know, the more you know that you don't know' is one that can certainly be applied to the behaviour of the domestic cat. On the surface, cats can seem to be simple, uncomplicated creatures that do little more than eat, sleep and take up room on our most comfortable chair or in front of the fire. But the more that you learn about their behaviour, the more interesting and complex they reveal themselves to be.

Historically, general interest and knowledge of the health, welfare and behaviour of our pet cats has taken second place behind that of our canine companions. That was certainly very true when I started out as a veterinary nurse more than 30 years ago. Even some 10 years or so later when I went on to study companion animal behaviour, the focus was still primarily on the behaviour of dogs. But things are changing; although there is still more scientific interest and research into canine behaviour, research into and resulting knowledge of feline behaviour is most definitely increasing and improving. However, in comparison to the number of books available on canine behaviour, there are still very few that concentrate solely on feline behaviour, especially those based on scientific research, and of those that do exist most are targeted primarily at students of companion animal behaviour and/or the veterinary professions. In fact, when this book was proposed, initially as a follow on from Stephanie Hedges' excellent 'Practical Canine Behaviour', the target audience was to be specifically veterinary nurses and technicians. But after some consideration and discussion with the wonderful people at CABI, it was decided to widen this to include anyone with an interest in feline behaviour, especially, but not exclusively, for those with a professional interest.

The book is divided into two parts. The first section should help to increase understanding of normal feline behaviour: the evolution of the domestic cat, how cats perceive the world around them, how they communicate, hunt, reproduce and learn, as well as how aspects such as physical health, stress and behaviour are closely linked.

The second section provides advice and information for specific groups of people involved with the care of cats: breeders, current and prospective cat owners, veterinary professionals, shelter and cattery workers, and others. But whatever heading you feel that you fit under, reading all these chapters should provide an overall picture of good feline care and behavioural welfare.

Acknowledgements

There are two very important things that I have learnt about writing a book such as this: one is how often you need to ask for help and advice from other people; and two is how incredibly helpful and generous people can be.

So my thanks go out to everyone who has helped me in this venture, most notably, in alphabetical order: Neil Andrews, Julie Bedford BSc(Hons) PGCE PGDipCABC CCAB, Dr. Sarah Ellis BSc(Hons) PGDipCABC PhD, Celia Haddon MA BSc MSc, Jackie Hart, Stephanie Hedges BSc (Hons) CCAB, Penny Stephens, Caroline Warnes BVSc MSc CCAB MRCVS and Clare Wilson MA VetMB CCAB MRCVS PGDipCABC.

Thank you also to the wonderful team at CABI, and everyone else (especially the Penguins) who have pointed me in the right direction to find a research paper, helped me when I couldn't think of a word or phrase a sentence, or just offered their thoughts and opinions.

Finally, much love and thanks my husband Alan, who has been, almost without complaint, a book-writing widower over the last 2 years but who has used the time constructively by honing his housework, cooking and ironing skills. And to Tui for offering constant companionship while I've been writing, and a fair bit of cat hair in my computer keyboard.

Part One
Understanding Feline Behaviour

Introduction

To be able to prevent behaviour problems and to ensure the best behavioural welfare for our pet cats, it is necessary to have some knowledge of normal behaviours and the factors that influence behaviour. Knowing what to do and how is, of course, essential but of no less importance is understanding why.

This section of the book aims to provide that 'why' information, as well as providing insights into the history, physiology and, of course, the behaviour of the domestic cat.

1 The Origin and Evolution of the Domestic Cat

There are approximately 40 different species of the cat family, classification *Felidae* (Table 1.1), all of which are descended from a leopard-like predator *Pseudaelurus* that existed in South-east Asia around 11 million years ago (O'Brien and Johnson, 2007). Other than the domestic cat, the most well known of the *Felidae* are the big cats such as lions, tigers and panthers, sub-classification *Panthera*. But the cat family also includes a large number of small cats, including a group commonly known as the wildcats, sub-classification *Felis silvestris* (Table 1.2).

Physical similarity suggests that the domestic cat (*Felis silvestris catus*) originally derived from one or more than one of these small wildcats. DNA examination shows that it is most closely related to the African wildcat (*Felis silvestris lybica*), which has almost identical DNA, indicating that the African wildcat is the domestic cat's primary ancestor (Lipinski *et al.*, 2008).

The African Wildcat

The African wildcat is still in existence today and is a solitary and highly territorial animal indigenous to areas of North Africa and the Near East, the region where domestication of the cat is believed to have first taken place (Driscoll *et al.*, 2007; Faure and Kitchener, 2009). It is primarily a nocturnal hunter that preys mainly on rodents but it will also eat insects, reptiles and other mammals including the young of small antelopes. Also known as the Arabian or North African wildcat, it is similar in appearance to a domestic tabby, with a striped grey/sandy-coloured coat, but is slightly larger and with longer legs (Fig. 1.1).

Felis lybica has long been considered a prime candidate as the ancestor of the domestic cat, even before the emergence of supporting DNA evidence, not only owing to its physical similarity and area of native origin, but also because it is comparatively less aggressive than other wildcats and it has been reported that attempts to tame and socialize African wildcat kittens with people have been successful. Other wildcats, although also similar in appearance to domestic cats, are not so tractable. The European wildcat (*Felis s. sylvestris*), also known as the Scottish wildcat, is particularly difficult to tame, and will remain fearful and aggressive even if hand raised and well socialized with people from an early age (Bradshaw, 2013; Serpell, 2014).

Table 1.1. Classification of *Felidae* (The Cat Family).

Lineage	Common Name	Species
Panthera		
	Lion	*Panthera leo*
	Leopard	*Panthera pardus*
	Jaguar	*Panthera onca*
	Tiger	*Panthera tigris*
	Snow Leopard	*Panthera uncia*
	Clouded Leopard	*Neofelis nebulosi*
	Sunda Clouded Leopard	*Neofelis diardi*
Bay Cat		
	Asiatic golden cat	*Catopuma temminckii*
	Bornean bay cat	*Catopuma badia*
	Marbled cat	*Pardofelis marmorata*
Caracal		
	Caracal	*Caracal caracal*
	African golden cat	*Caracal aurata*
	Serval	*Leptailurus serval*
Ocelot		
	Geoffroys's cat	*Leopardus geoffroyi*
	Guiña, Kodkod	*Leopardus guigna*
	Northern tiger cat	*Leopardus tigrinus*
	Southern tiger cat	*Leopardus guttulus*
	Andean cat	*Leopardus jacobita*
	Margay	*Leopardus wiedii*
	Pampas cat	*Leopardus colocolo*
	Ocelot	*Leopardus pardalis*
Lynx		
	Iberian lynx	*Lynx pardinus*
	Eurasian lynx	*Lynx lynx*
	Canada lynx	*Lynx Canadensis*
	Bobcat	*Lynx rufus*
Puma		
	Puma	*Puma concolor*
	Cheetah	*Acinonyx jubatus*
	Jaguarundi	*Herpailurus yagouaoundi*
Leopard Cat		
	Leopard cat	*Prionailurus bengalensis*
	Fishing cat	*Prionailurus viserrinus*
	Flat-headed cat	*Prionailurus planiceps*
	Rusty-spotted cat	*Prionailurus rubiginosus*
	Pallas's cat	*Octocolbus manul*
Felis		
	Wildcat	*Felis silvestris*
	Sand cat	*Felis margarita*
	Black-footed cat	*Felis nigripes*
	Jungle cat	*Felis Chaus*

Table 1.2. Sub-species of small wildcats (*Silvestris* spp.).

Common name	Species
European (Scottish) wildcat	*Felis silvestris silvestris*
Indian desert cat	*Felis silvestris ornata*
Chinese mountain cat	*Felis silvestris bieti*
South African wildcat (Caffer cat, Bush cat)	*Felis silvestris cafra*
African wildcat (aka North African or Arabian wildcat)	*Felis silvestris lybica*
Domestic cat	*Felis silvestris catus*

Fig. 1.1. The African wildcat (*Felis silvestris lybica*). The most likely ancestor of the domestic cat.

Domestication

Definition

There is a distinct difference between an animal that has been tamed and a domesticated species.

- **A tamed animal** is an individual from a wild population that has been conditioned to no longer fear people. There is no human influence on other members of the same species or on the genetics of the species.
- **A domesticated species** refers to a taxonomic group of animals whose behaviour, physiology and genetics have been altered by selective breeding. Domestication is a gradual process that can take several generations.

The beginnings of feline domestication

The domestication of the cat probably began around 10,000 years ago in a region of the Middle East known as the Fertile Crescent (Fig. 1.2) (Vigne *et al.*, 2004; Driscoll

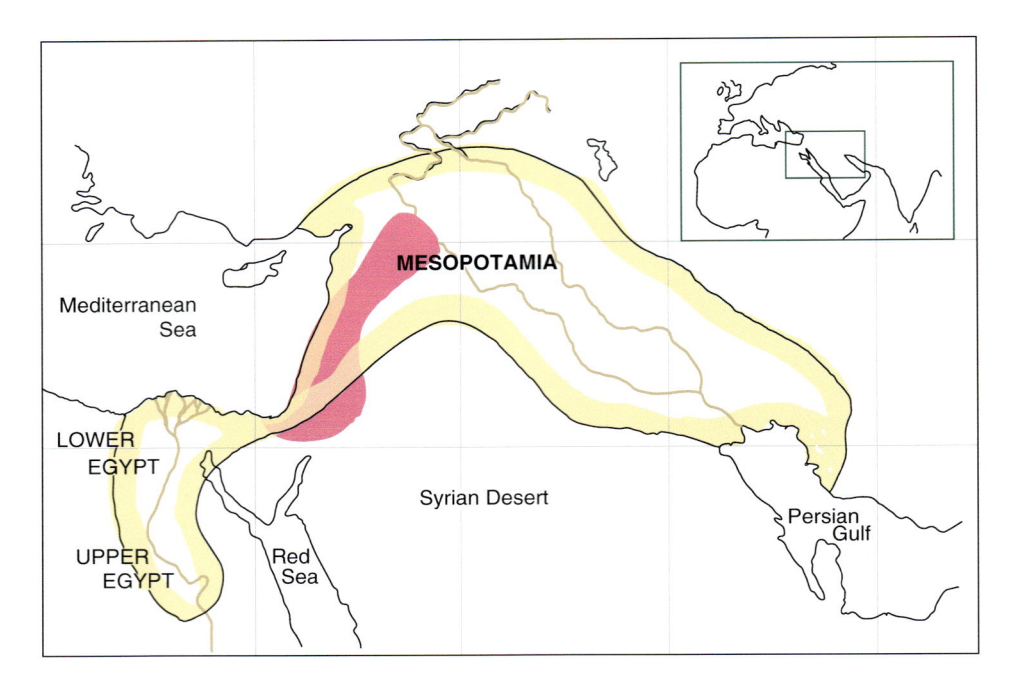

Fig. 1.2. The Fertile Crescent, the area believed to be where the domestication of the cat began. Pink denotes the area inhabited by the Natufian people 12,800 to 10,300 years ago, the first people to build permanent stone houses and believed to be the originators of agriculture.

et al., 2007; Lipinski *et al.*, 2008; Bradshaw, 2013). This area is so called because it is thought to be the birthplace of farming, which occurred during the Natufian period 12,800 to 10,300 years ago. The Natufian people were the first to build permanent stone houses and are believed to be the originators of agriculture (Bar-Yosef, 1998).

It would have been necessary to store crops produced by farming to provide food throughout the year. Grain and other food stores would have attracted vermin, and it is believed that this played a large part in the evolution and eventual widespread multiplicity of the house mouse (*Mus musculus*). These and other rodents, including rats, were likely to have become present in large numbers around human settlements. This is supported by archaeological discoveries of pottery rodent-traps (Filer, 2003). But another means of rodent control was also likely to have presented itself in the form of natural predators, including the African wildcat, which would have been attracted to human habitation by the high predominance of and easy access to prey.

It is speculated that the domestication of the cat took place by a variety of means:

- The cats that were better able to cope with being near human settlements would have had increased hunting opportunities and so better survival prospects. It is therefore theorized that cats might have 'domesticated themselves' by increased association with humans, which would have selected for calmer and more tameable individuals (Leyhausen, 1988).
- Because of their usefulness as vermin controllers it is also likely that cats were fed and encouraged by people to stay close to human settlements.

- The domestication process might have been further assisted by people capturing, adopting and taming kittens, a process that still occurs today in some primitive societies such as Amazonian tribes, who capture and make pets of various jungle species (Serpell, 2014).

Ancient Egypt – the hub of domestication

The beginnings of domestication may have taken place in other areas of the Near and Middle East but most evidence of large-scale domestication comes from ancient Egypt.

There are many ancient Egyptian pictorial and hieroglyphic references to cats. In the earliest illustrations and hieroglyphs there was rarely any lexicographical discrimination between wild or domestic cats, or between the different species of wildcats. Therefore, it is often unclear if the cats depicted were wild or domesticated (Malek, 1993). During the Middle Kingdom period (2025–1606 BC), however, cats began to be depicted in association with people and a set of hieroglyphs translated as 'miw' start to appear indicating the domestic cat. Later, during the New Kingdom period (1539–1070 BC) illustrations of cats in domestic situations increased to where they appear to have become an accepted part of everyday life (Serpell, 1988; Filer, 1995; Bradshaw, 2013).

Historical Attitudes Towards Cats

Human attitudes towards cats have varied tremendously throughout history, from being highly revered in ancient Egypt to being subjected to widespread hatred and persecution throughout much of Europe from the Middle Ages and beyond.

The Cat in the Ancient Egyptian religion

Animals played an important part in Egyptian religion. The Egyptians did not, as is commonly thought, worship the animal itself but they believed that the gods could manifest themselves as animals; they therefore considered members of a species associated with a god or goddess to be representations of, and be imbued with, the spirit of that deity. The cat had other godly associations but became particularly associated with the goddess Bastet. Bastet represented fertility, child bearing, general care, nurturing and protection, which made her a very popular goddess with the masses who suffered high levels of child mortality and a generally high death rate. Her role as a nurturing mother figure also made her a deity of choice for anyone seeking help and comfort.

European attitudes to cats from the Middle Ages and beyond

Attitudes to cats in early medieval Western Europe appear to have been positive or at least benign. They were highly valued for their rodent-killing ability and although English bishops tried to ban other animals being kept in nunneries and monasteries, they allowed, even encouraged, the keeping of cats, from where there are also records that they were occasionally kept as fond companions (Newman, 1992). Celtic monks also included cats favourably in their illustrations. By the 13th century, however, cats were not viewed so favourably by the Christian church, most possibly because of their association with some of the pre-Christian religions. At around this time, a widespread hatred and persecution developed that persisted for over 400 years (Engels, 1999; Lockwood, 2005).

Companion animals, but most especially cats, also came to be associated with witches and witchcraft. This association might have evolved from a fear and condemnation of pagan religions that involved the worship of female deities, most notably those for whom the cat was a divine incarnation such as Bastet, Isis, Artemis and Diana (Engels, 1999).

The persecution of cats, sanctioned by the church, became so widespread that they were often killed for 'good luck' and in many European areas their mass slaughter, usually by burning, became an accepted part of some feast days and festivals (Lockwood, 2005). Some of these ritual 'celebrations' persisted into modern times, especially in some areas of rural France where they continued well into the 18th century (Darnton, 1984; cited in Lockwood, 2005.)

The Modern-day Pet Cat

Pedigree cat breeds

Widespread selective breeding of cats to produce distinct breeds is a fairly recent activity that began in the late 19th century. When the first recorded cat show was held at the Crystal Palace in London in 1871 the cats were separated into two groups: long-haired and short-haired with four 'types' in the long-haired group and 12 in the short-haired section (Weir, 1889). Many of these were defined by no more than coat colour so they might have been just naturally occurring variations, rather than as a result of selective breeding.

In comparison, The Governing Council of the Cat Fancy (GCCF), the modern-day registry for pedigree cats in the UK, recognizes more than 65 breeds, separated into seven groups. And in the USA, where there are two main feline breed registries, 43 different breeds are recognized by the Cat Fanciers Association (CFA) and around 70 by The International Cat Association (TICA).

Behavioural and personality traits of cat breeds

When reading about individual cat breeds, information about behavioural traits is usually supplied as well as a description of the breed's physical appearance. But there has been very little scientific research into breed-specific behaviour in cats.

One major difference between the selective breeding of dogs and that of cats is that dogs were originally selectively bred to enhance behavioural as well as physical characteristics, whereas cats have generally been bred to enhance physical differences only. Therefore breed-related behavioural differences would be expected to occur far more in dogs than in cats. Even so, research into dog behaviour has shown that, even with dogs, there is often as much behavioural variation between individuals as there is between different breeds and this can certainly be the case with cats as well.

What can have an indirect influence on behaviour, however, are the extreme changes from the normal physical shape and a predisposition towards inherited disorders that can result from selective breeding. These might cause life-long pain and discomfort that in turn might negatively influence behaviour, as well as being a serious health and welfare concern (Fig. 1.3 a–c).

Domestic/wild hybrids

In recent years there has been an increasing fashion for hybrid cat breeds (Table 1.3). There are insufficient data available to allow accurate advice to be given as to the expected behaviour or potential behaviour problems associated with these breeds. However, the fact that they have genetic input from wild felines, some of which are

Fig. 1.3. (a) Munchkins are an extreme example of how selective breeding from mutations and deformities can produce animals that are very different from the normal feline physical shape. The short legs and long spine of the Munchkin can restrict the cat's movement and the ability to groom itself, plus the development of abnormal joints can be painful and debilitating. The Munchkin is not recognized by the Governing Council of the Cat Fancy (GCCF) in the UK or by the Cat Fanciers Association (CFA). (b) Another example of extreme and detrimental change from the natural feline shape is the brachycephalic (short nose) Persian. These cats can suffer from a wide range of health problems including: corneal sequestrum development and persistent ocular discharge owing to abnormally large protruding eyes; dental disease owing to a shortened jaw and misaligned teeth; and of most concern severe breathing difficulties owing to a severely reduced nasal opening, narrowed nasal passages and often an over-long soft palate. (c) The Scottish Fold is a breed whose physical appearance is linked to a debilitating genetic mutation that affects the development of cartilage. This mutation (osteochondrodysplasia) not only affects the ears, giving the cat its characteristic 'folded' ears, but also the cartilage and bone in the joints leading to severe and painful arthritis.

Table 1.3. Hybrid cat breeds.

Breed	Genetic input
Bengal	Asian leopard cat (*Prionailurus bengalensis*)
Chausie	Jungle cat (*Felis chaus*)
Savannah	Serval (*Leptailurus serval*)
Safari	Geoffroys's cat (*Leopardus geoffroyi*)

known to be difficult if not impossible to tame, raises concerns regarding the safety of keeping these cats in a home environment and for the welfare of the cats themselves.

Current Attitudes Towards Cats and Keeping Cats as Pets

Sadly, a professed hatred or mistrust of cats is still not uncommon, even in countries and cultures where there is a high level of pet cat ownership. Cats are also more likely to be subjected to intentional physical abuse than dogs (Lockwood, 2005). Research has found, however, that the attitude of most people towards cats is now generally positive (Turner, 2014).

In many developing countries, cats are still kept primarily for vermin control, but the keeping of cats for companionship is a growing practice worldwide and is more popular now than it has ever been, especially in areas of greater economic growth and urbanization (Bernstein, 2007). In the UK, approximately 40% of all households own a pet with around 17% owning at least one pet cat (PFMA, 2017). Over the whole of Europe, excluding Russia, there are around 75 million pet-owning households, with cats surpassing dogs as the most popular (FEDIAF, 2014). In the USA more than half of all households own at least one pet animal, an estimated 30% being cats (AVMA, 2012).

Why do we keep cats as pets?

Although every pet cat owner will have their own reasons why they have a cat, for many the initial appeal can be the cat or kitten's appearance. Their round faces, large, forward-facing eyes and small noses give cats a look reminiscent of a human infant, thereby appearing 'cute' and engendering nurturing feelings in human adults, especially women. The type of pet that someone chooses can also be highly influenced by previous experience. People that have owned cats previously, especially if they grew up with them are more likely to keep cats as adults (Serpell, 1981). The cat's independence can also be an attraction for some, although this is more likely to be the case for owners that allow their cats outdoor access. Owners of indoor-only cats are less likely to consider this to be a desirable characteristic (Turner, 1995).

Attachment

An attachment can be described as a close emotional bond with another individual. Most pet owners, especially cat and dog owners, develop an emotional attachment to their pet similar to that experienced with close friends and family members.

Studies have also shown that there are physiological and psychological health benefits to be gained from pet ownership (Friedmann and Thomas, 1995; Bernstein, 2007; Kanat-Maymon et al., 2016). Stroking or just being near to a cat that an individual has an attachment to has been shown to effectively reduce heart rate and lower blood pressure (Dinis and Martins, 2016).

Pets can also provide owners with emotional support and for people who live alone or have limited social interactions, a dog or cat might be the only consistent emotional support available (Stammbach and Turner, 1999).

Feline Behaviour

Despite the popularity of cats as pets, the overall level of knowledge in feline behaviour appears limited and nowhere near as widespread as the increasing general awareness of canine behaviour (Riccomini, 2009; Pereira et al., 2014). Sadly, a lack or misunderstanding of feline behaviour by well-meaning and often compassionate cat owners and carers can be a major contributory factor in the development of feline stress and associated health and behaviour conditions (see Chapter 6).

Anthropomorphism

Anthropomorphism is the attributing of human characteristics and thoughts to non-human animals. Many pet owners exhibit some degree of anthropomorphism when talking about or interacting with their pets. But as long as they are aware that in reality their pets are unlikely to possess these human traits and they do not subject the pet to 'human like' activity that may cause the animal distress, this level of anthropomorphism can be a normal part of owner attachment to the pet and a healthy pet–owner relationship. If the owner has very little understanding of species-specific behaviour, however, the human perception might be their only point of reference. When this is the case it can present welfare concerns for the animal, especially if the owner or carer has a strong and resolute belief that their pets are 'small people' and that the animal's physical and behavioural needs are the same as that of a human or human child.

Feline behaviour problems

Behaviour problems can be unpleasant for the cat owner and are often a sign of feline distress. The existence of behaviour problems in the pet cat population is far from uncommon and yet cat owners are much less likely than dog owners to seek professional help and advice (Bradshaw et al., 2000). This might be due to the following factors:

- The behaviour of pet dogs can affect other people and might even lead to prosecution for the owner. A cat's behaviour is, however, less likely to be of concern to anyone other than the owner, which puts far less obligation on the cat owner to address their pet's behaviour.
- Unless the behaviour is one that is directly unacceptable to the owner, e.g. house-soiling or human-directed aggression, many issues go unrecognized or can be incorrectly regarded as being a part of normal cat behaviour.

- Stress can have harmful effects on feline health, behaviour and welfare. In one study over 70% of cat owners questioned were aware of this, but the majority of those questioned were unable to recognize the signs of feline stress and related health issues (Mariti et al., 2015).
- Even if owners regard their cat's behaviour as unacceptable or of concern, it is a common mistaken belief that behaviour issues can only be addressed by training. When accompanied with the incorrect conviction that 'cats can't be trained', the perception is that the treatment of feline behaviour problems is therefore impossible.
- But even when none of the above applies, professional help might still not be sought owing to a lack of awareness that help is available and/or who to approach for advice and assistance (Riccomini, 2009). See Chapter 11b for information and advice regarding help for behaviour problems.

References

AVMA (2012) U.S. Pet Ownership Statistics. Available at: https://www.avma.org/KB/Resources/Statistics/Pages/Market-research-statistics-US-pet-ownership.aspx (accessed 12 January 2018).

Bar-Yosef, O. (1998) The Natufian culture in the Levant. *Evolutionary Anthropology* 6, 167–168.

Bernstein, P.L. (2007) The human–cat relationship. In: Rochlitz, I. (ed.) *The Welfare of Cats*. Springer, Dordrecht, The Netherlands.

Bradshaw, J.W.S. (2013) *Cat Sense: The Feline Enigma Revealed*. Allen Lane, Penguin Books, London.

Bradshaw, J.W.S., Casey, R.A. and MacDonald J.M. (2000) The occurrence of unwanted behaviour in the pet cat population. *Proceedings of the Companion Animal Behaviour Therapy Study Group Study Day*, Birmingham, UK, pp. 41–42.

Darnton, R. (1984) *The Great Cat Massacre*. Basic Books, New York. Cited in: Lockwood, R. (2005) Cruelty to cats: changing perspectives. In: Salem, D.J. and Rowan, A.N. (eds) *The State of the Animals III: 2005*. pp. 15–26, Humane Society Press, Washington, DC.

Dinis, F.A. and Martins, T.L.F. (2016) Does cat attachment have an effect on human health? A comparison between owners and volunteers. *Pet Behaviour Science* 1, 1–12.

Driscoll, C.A., Menotti-Raymond, M., Roca, A.L. et al. (2007) The Near Eastern origin of cat domestication. *Science* 317, 519–523.

Engels D. (1999) *Classical Cats. The Rise and Fall of the Sacred Cat*. Routledge, London.

Faure, E. and Kitchener, A.C. (2009) An archaeological and historical review of the relationships between felids and people. *Anthrozoös* 22, 221–238.

FEDIAF (2014) European Pet Food Industry Federation Statistics 2014. Brussels. Available at: www.fediaf.org (accessed 12 January 2018).

Filer, J. (1995) Attitudes to death with reference to cats in ancient Egypt. In: Campbell, S. and Green, A. (eds) *The Archaeology of Death in the Ancient Near East*. Oxbow Books, Oxford, UK.

Filer, J. (2003) 'Without exception, held to be sacred': The Egyptian cat. Ancient Egypt. *The History, People and Culture of the Nile Valley* 4(2), 24–29.

Friedmann, E. and Thomas, S.A. (1995) Pet ownership, social support and one year survival after acute myocardial infarction in the Cardiac Arrthymia Suppression Trial (CAST). *American Journal of Cardiology* 76, 1213–1217.

Kanat-Maymon, Y., Antebi, A. and Zilcha-Mano, S. (2016) Basic psychological need fulfilment in human-pet relationships and well-being. *Personality and Individual Differences* 92, 69–73.

Leyhausen, P. (1988) The tame and the wild – another Just-so Story? In: Turner, D.C. and Bateson, P. (eds) *The Domestic Cat: The Biology of its Behaviour*, 1st edn. Cambridge University Press, Cambridge, UK, pp. 57–66.

Lipinski, M.J., Froenicke, L., Baysac, K.C. *et al.* (2008) The ascent of cat breeds: Genetic evaluations of breeds and worldwide random-bred populations. *Science Direct* 91(1), 12–21.

Lockwood, R. (2005) Cruelty to cats: changing perpectives. In: Salem, D.J. and Rowan, A.N. (eds) *The State of the Animals III: 2005*, pp. 15–26, Humane Society Press, Washington, DC.

Malek J. (1993) *The Cat in Ancient Egypt*. British Museum Press, London.

Mariti, C., Guerrini, F., Vallini, V., Guardini, G., Bowen, J., Fatjò, J. and Guzzano, A. (2015) The perception of stress in cat owners. In: *Proceedings of the 2015 AWSELVA-ECAWBM-ESVCA Congress*, Bristol, UK.

Newman, B. (1992) The Cattes Tale: a Chaucer apocryphon. *The Chaucer Review* 26, 411–423.

O'Brien, S. J. and Johnson W.E. (2007) The evolution of cats. *Scientific American* 297, 68–75. DOI: 10.1038/scientificamerican0707-68.

Pereira, G.D.G., Fragoso, S., Morais, D., Villa de Brito, M.T. and de Sousa, L. (2014) Comparison of interpretation of cat's behavioural needs between veterinarians, veterinary nurses and cat owners. *Journal of Veterinary Behavior* 9, 324–328.

PFMA (2017) Pet Food Manufacturers Association – Statistics 2016. Available at: http://www.pfma.org.uk/historical-pet-population (accessed 12 January 2018).

Riccomini, F.D. (2009) The cat owner – feline friend or foe? *Proceedings of the British Veterinary Behaviour Association*. Birmingham, UK.

Serpell, J.A. (1981) Childhood pets and their influence on adults' attitudes. *Psychological Reports* 49, 651–654.

Serpell, J.A. (1988) The domestication and history of the cat. In: Turner, D.C. and Bateson, P. (eds) *The Domestic Cat: The Biology of its Behaviour,* 1st edn. Cambridge University Press, Cambridge, UK.

Serpell, J.A. (2014) Domestication and history of the cat. In: Turner, D.C. and Bateson, P. (eds) *The Domestic Cat: The Biology of its Behaviour,* 3rd edn. Cambridge University Press, Cambridge, UK.

Stammbach, K.B. and Turner, D.C. (1999) Understanding the human–cat relationship: human social support or attachment. *Anthrozoös* 12, 162–168.

Turner, D.C. (1995) The human–cat relationship. In: Robinson, I. (ed.) *The Waltham Book of Human-Animal Interaction: Benefits and Responsibilities of Pet Ownership.* Waltham Centre for Pet Nutrition, Elsevier, London, pp. 87–97.

Turner, D.C. (2014) Social organisation and behavioural ecology of free-ranging domestic cats. In: Turner, D.C. and Bateson, P. (eds) *The Domestic Cat: The Biology of its Behaviour,* 3rd edn. Cambridge University Press, Cambridge, UK.

Vigne, J-D., Guilaine, J., Debue, K., Haye, L. and Gérard, P. (2004) Early taming of the cat in Cyprus. *Science* 304, 259.

Weir, H. (1889) *Our Cats and All About Them*. R. Clements and Co., Tunbridge Wells, UK.

2 The Senses

Because pet cats share our lives and our homes it can be easy to assume that their experience and perception of surrounding stimuli is the same as ours. But their sensory abilities, which have not changed from their wild ancestry, enable them to see, hear and smell things of which our senses leave us completely unaware. In a few respects, however, our senses are slightly better or at least just dissimilar but the overall effect is that the cat's view of the world is quite different to our own.

Sight

Night vision

There are a number of ways in which a domestic cat's vision differs from human eyesight. The most well-known of these is the cat's ability to 'see in the dark'. In truth cats cannot see any better than we can in complete darkness but in low light conditions specific structures of the feline eye allow them to utilize any available light far better than we can.

- The retina of mammalian eyes contains two types of photoreceptors: rods and cones. Cones function best in bright light and are responsible for colour vision. Rods are more light sensitive and function better in reduced light to allow vision in low-light conditions. A cat's retina contains far more rods than cones, with the total number of rods being around three times greater than is found in the human eye (Steinberg et al., 1973).
- The rod photoreceptor cells are not connected to fibres in the optic nerve individually but are joined together in bundles, so that more visual receptors are connected to each nerve cell, resulting in even greater light detection. The disadvantage of this, however, is that it also reduces the clarity of the image (Miller, 2001).
- Cats have very large eyes in relation to body size. At 22 mm in diameter, they are only slightly smaller than human eyes, which are 25 mm. The large size of the eyes along with elliptical pupils allow for pupil dilation that is at least three times greater than our own, enabling a large amount of light to enter the eyes when required. In bright light the elliptical pupil can be reduced by constriction of the iris to a very small vertical slit only 1 mm wide to protect the retina from damage. (Fig. 2.1).
- A layer of reflective cells behind the retina called the 'tapetum lucidum' – more specifically the tapetum cellulosum in cats and some other carnivores – reflects light that has not yet been absorbed by the photoreceptor cells back into the eye to allow another chance for absorption. This increases the efficiency of the eye in low light conditions by up to 40% (Bradshaw et al., 2012). The tapetum is responsible for the characteristic reflective shine whenever a light is shone into a cat's eyes.

Fig. 2.1. Elliptical pupils can dilate more than round pupils, allowing more light to enter the eye. They can also constrict to very narrow slits to avoid damage to the retina from bright light.

Movement detection

Rapid eye movements known as 'saccades' prevent moving images from blurring and enable cats to detect and accurately track fast movements, even if the movement is slight or to the side of the cat's direct line of vision. Sensory information regarding the distance and angle of the object is sent to the muscles surrounding the lens (the ciliary muscles) but, if this calculation is incorrect or if the object moves in an unexpected way, a second corrective saccade follows the first to get the line of vision back on target. This corrective imaging can occur around 60 times per second, at least twice as fast as we are capable of (Bradshaw *et al.*, 2012).

Slow movements, however, are not so well detected. Humans can detect movements that are ten times slower than those that can be seen by cats (Pasternak and Merigan, 1980).

Visual focusing

Visual focusing is one way in which our eyes perform much better than those of cats. In human eyes, the lens is flexible and, depending on the overall health of the eye, is able to focus fairly accurately by distortion of the lens. But the lens within a cat's eye is inflexible and focuses by moving backward and forward rather than distorting. The result is that cats are a little slower at transferring their visual focus from near to distance and vice versa and are unable to focus clearly on anything that is less than 25 cm away from their face.

Colour vision

Perception of colour is another way that human vision is superior. Humans have three types of colour sensitive cone photoreceptors (red, blue and yellow) and 16 times more colour-comparing nerves than cats, which allow us to see a far wider range of different colours.

Rod photoreceptors only allow for monochromatic vision and it was once thought that cats can only see in black and white. But neurophysiological evidence shows that they possess two types of colour-sensitive cones that respond to light within the blue and green light spectrum, which means that they are likely to see yellow and blue as primary colours and their combination colours (Loop *et al.*, 1987). Behavioural research indicates, however, that cats' colour perception may be muted or that they are 'behaviourally colour blind' because they seem much more able to distinguish between differences in shape, pattern and brightness than between colours (Bradshaw *et al.*, 2012).

Binocular vision

Like us, cats have forward facing eyes that work together giving them stereoscopic or 3D vision (DeAngelis, 2000). This is achieved by each eye focusing on the same visual target and producing its own 2D image. These images overlap with each other resulting in 3D vision. Stereoscopic vision is a great advantage for a predator because it allows for accurate judgment of depth and distance.

Some Siamese cats have a genetic misrouting of retinal ganglion cells resulting in an inability to develop full stereoscopic vision (Bacon *et al.*, 1999). However, other than the development of an adaptational squint or 'cross-eyes', these cats appear otherwise unaffected and many are perfectly able to hunt successfully. Cross-eyed Siamese were quite common at one time, but are not seen so often now as more breeders have become aware of the genetic influence and choose not to breed from affected cats.

Field of vision

The field of vison is the area that can be seen when the eyes are fixed in one position. It is made up of the binocular visual field, where the two eyes work together to provide a stereoscopic image, and the peripheral field which is outside of the central gaze and is non-binocular. Cats have about the same binocular field of vision as humans, about 90–100°. But their lateral peripheral visual field is wider, giving them a total of around 200° compared to around 180° in humans.

Hearing

Cats have one of the widest ranges of hearing among mammals, extending from 48 Hz to 85 kHz (Heffner and Heffner, 1985), although it is generally accepted that the useful upper limit is more likely to be around 60 kHz because sounds above this frequency would have to be fairly loud for a cat to be able to hear them (Bradshaw *et al.*, 2012). Even so, it is still quite exceptional when you consider that the average human hearing range is approximately 20 Hz to 20 kHz.

The hearing range of most mammals encompasses that necessary to hear species-specific auditory communications plus, if a predatory species, the sounds made by prey. Hearing is one of the cat's main methods of prey detection; it is therefore not surprising that their hearing range enables them to hear the very high-pitched 'ultrasound' signalling made by small rodents and other small prey species. But their hearing range also encompasses much lower frequency sounds than would be expected. This can be a particular advantage for pet cats because it allows them to hear the full range of human speech, even the low-pitched human male voice. Cats also have very sensitive hearing, allowing them to hear not just the vocalizations but also the movements of small animals and insects.

Their ability to detect minor differences in pitch and intensity between sounds is, however, inferior to our own and they are less able to detect sounds of very short duration. Our increased ability in this respect might be due to our need to detect such small changes in sound that occur in human language (Bradshaw, 2013).

The pinna (the external part of the ear) plays a very important role in both the location of sounds and as a directional amplifier. Individual muscles allow each pinna to be moved independently and these movements can be both rapid and precise enabling the cat to accurately pinpoint the source of a sound.

Olfaction (Sense of Smell)

Olfaction is an extremely important sense for cats because it is used in communication, reproduction, feeding and hunting. It is therefore not surprising that cats have a highly sensitive and discriminative sense of smell.

Olfactory receptor cells, which have a direct neural connection to the olfactory bulb in the brain, are found in the olfactory epithelium – the lining of the nasal cavities. In humans this covers an area of around 2–5 cm^2 and contains around 5 million receptors. In cats the olfactory epithelium is supported by the scroll-like nasal turbinate bones, the ethmoturbinals, and covers a total surface area of approximately 20–40 cm^2, containing around 200 million receptors (Ley, 2016). These are receptors of not just one type, but several hundred different types, enabling the cat to distinguish between an incredibly enormous number of different odours (Bradshaw *et al.*, 2012).

The vomeronasal organ

The vomeronasal organ, also known as the Jacobson's organ, is an additional olfactory organ found in many mammals but not in humans or other higher primates. It provides the animal with a sense that is probably something between taste and smell and appears to be used primarily to detect and identify pheromones used in intra-species communication (see Chapter 3). It consists of two blind-ended fluid-filled sacs, situated within the hard palate, that contain around 30 different types of chemical receptors. Scent is actively taken in through the mouth and towards the vomeronasal organ via two small slits, known as the nasopalatine ducts, positioned just behind the upper incisors.

When utilizing the vomeronasal organ, a cat will exhibit a slightly odd facial expression or 'grimace' whereby the mouth is held slightly open with the upper lip raised (Fig. 2.2). This is also known as the flehmen response, from the German verb meaning to bare the upper teeth. Flehmen is a voluntary behaviour and it is possible that the presence of other visual or olfactory signals, such as the scent of urine, might trigger this reaction (Mills *et al.*, 2013).

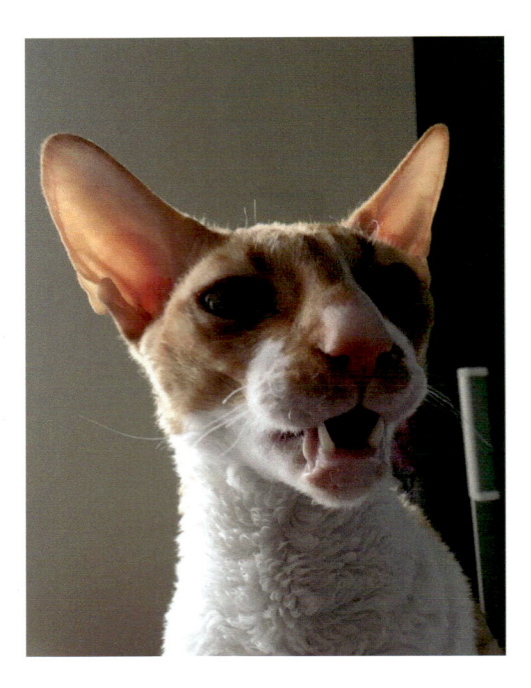

Fig. 2.2. The 'flehmen response', activating the vomeronasal organ. © Lucy Hoile.

Touch

The vibrissae (whiskers)

These are:

- The mystacials: the large whiskers on either side of the nose.
- The supercilliary: situated just above the eyes
- The genals: smaller hairs situated on the side of the face.
- Carpel hairs: found on the back of the forelegs.

The whiskers, scientifically known as the vibrissae, are thickened hairs embedded deeply within the skin, around three times deeper than regular guard hairs. Large numbers of mechanoreceptors, sensory neurons responsive to pressure or distortion, are found at the base of the vibrissae, making them sensitive enough to detect air currents. When the cat is hunting, the large facial whiskers (the mystacials) take over from the eyes to compensate for the cat's poor close visual focusing. This is achieved by the whiskers being pushed forward to envelope the prey when it is close to the face enabling the cat to accurately detect and manipulate it. Cats are also able to get the best overall picture of their close surroundings and position in relation to nearby obstacles by combining visual input with sensory information from the whiskers (Bradshaw *et al.*, 2012).

The feet

A cat's feet play an important role in hunting and defence, and are used to investigate and explore novel objects by touch. They are highly sensitive, containing a very high density of mechanoreceptors both in and between the pads and in the soft tissue at the base of the claws. Specialist mechanoreceptors known as Pacinian or Lamellar corpuscles in the deeper skin layers also allow cats to detect vibrations through the pads of the feet (Verrillo, 1966).

Balance

The vestibular system

The vestibular system is the part of the inner ear responsible for the sense of balance in mammals. It is made up of the semicircular canals, and the saccular and utricular otolith organs.

The semicircular canals are three fluid-filled tubes that also contain motion-sensitive hairs (cilia). They are:

- **The Horizontal:** which detects turns to the left or right.
- **The Anterior and Posterior:** which detect up and down movement and movements such as putting the head to one side.

As the animal moves its head the vestibular system moves along with it but the fluid within the semicircular canals stays in place, thereby moving the hairs within the canals. The information about the movement is then relayed to the brain.

The saccular and utricular otolith organs also have cilia covering their internal surface, but rather than fluid these organs have tiny calcite crystals covering the sensory surface that brush against the hairs as the animal moves its head. The otolith organs detect acceleration, deceleration and gravity, allowing the animal to know when and how fast it is moving, and when it is the 'right way up'.

Although this is the same system as in other mammals, including humans, the cat's semicircular canals are much nearer to being at right angles to each other than in many other mammals and the horizontal canal is more parallel to the normal head position. This allows the information detected by the cilia within the canals and relayed to the brain to be particularly clear and precise. Plus, the utricular otoliths, and possibly the saccular, are much better attuned to measure gravitational deviations (Bradshaw *et al.*, 2012).

The 'Righting' reflex

The 'righting reflex,' sometimes called the 'air-righting reflex' is the cat's ability to land on its feet from a fall. When a cat starts to fall the movement is instantly detected by the vestibular system and within one tenth of a second the head starts to turn towards the ground, allowing the cat to see where it will land. The highly flexible spine allows the body to then twist, front end first and then the back end so that the cat is facing the ground. Lastly, as the cat is about the land, the back arches and the legs extend ready to act as shock absorbers (Fig. 2.3).

If the fall is from a considerable height the legs are initially pushed out sideways and only extended downwards as the cat is just about to land. This has the effect of reducing the falling speed and probably accounts for reports of cats surviving falls

Fig. 2.3. The righting reflex. As soon as the cat starts to fall the movement is detected by the vestibular system and within one tenth of a second the body starts to twist to allow the cat to land on its feet.

from tall buildings with only minor injuries. But this does not always prevent cats from sustaining serious injury as it can also depend on the nature of the surface on which the cat lands. Injuries can also occur if the distance that the cat falls is too short. Although the righting reflex happens very quickly, a distance of at least 10 feet is needed to allow the cat to completely right itself (Bradshaw, 2013).

References

Bacon, B.A., Lepore F. and Guillemot, J-P. (1999) Binocular interactions and spatial disparity sensitivity in the superior colliculus of the Siamese cat. *Experimental Brain Research* 124, 181–192.

Bradshaw, J.W.S. (2013) *Cat Sense*: *The Feline Enigma Revealed*. Allen Lane, Penguin Books, London.

Bradshaw, J.W.S., Casey, R.A. and Brown, S.L. (2012) *The Behaviour of the Domestic Cat*, 2nd edn. CAB International, Wallingford, UK.

DeAngelis, G.C. (2000) Seeing in three dimensions: the neurophysiology of stereopsis. *Trends in Cognitive Science* 4, 80–90.

Heffner, S. and Heffner, H.E. (1985) Hearing range of the domestic cat. *Hearing Research* 19, 85–88.

Ley, J. (2016) Feline communication. In: Rodan, I. and Heath, S. (eds) *Feline Behavioral Health and Welfare*. Elsevier, St Louis, Missouri, USA.

Loop, M.S., Millican, C.L. and Thomas, S.R. (1987) Photopic spectral sensitivity of the cat. *Journal of Physiology* 382, 537–553.

Miller, P.E. (2001) Vision in animals – what do dogs and cats see? *Waltham/OSU Symposium. Small Animal Opthalmology*. Ohio State University, Waltham, Ohio, USA.

Mills, D., Dube, M.B. and Zulch, H. (2013) *Stress and Pheromonatherapy in Small Animal Clinical Behaviour*. John Wiley & Sons, Ltd, Oxford, UK.

Pasternak, T. and Merigan, W.H. (1980) Movement detection by cats: invariance with direction and target configuration. *Journal of Comparative and Physiological Psychology* 94, 943–952.

Steinberg, R.H., Reid, M. and Lacy, P.L. (1973) The distribution of rods and cones in the retina of the cat (*Felis domesticus*). *Journal of Comparative Neurology* 148, 229–248.

Verrillo, R.T. (1966) Vibrotactile sensitivity and the frequency response of the Pancinian corpuscle. *Psychonomic Science* 4, 135–136.

3 Feline Communication

Communication is the transfer of information from one individual to another. It can be intra-specific, meaning between members of the same species, or inter-specific, from one species to another. Human language is a highly complex form of intra-specific communication that enables us to exchange a wide range of information about ourselves and knowledge that we have gathered from experience or received from others. But communication between members of other species, including the cat, is generally limited to defensive or offensive warnings, greetings, care or food soliciting and basic information about the individual such as sexual status and fitness. To convey these signals a variety of different methods are used, including auditory (vocalizing), visual (body 'language' and facial expressions) and olfactory (scent). There are advantages and disadvantages in the use of each of these (Table 3.1), and a combination of methods may be employed. The method used can also depend on the intended signal and/or recipient.

The ancestor of the domestic cat, the African wildcat (*Felis s. lybica*) is a solitary animal, so the signalling repertoire between adults is primarily territorial. But domestication has resulted in a number of social and behavioural changes, including the ability to live in social groups and with other species. So, although the domestic cat has retained its ancestor's territorial signals and means of communication most suited for a solitary species, it has also developed its communication skills to suit its needs as a more social species.

Vocalizations

The sounds made by cats can be separated into three groups.

Vowel sounds

These are produced by the cat opening and closing its mouth and include:

- The meow.
- The sexual vocalizations: the female call when in oestrus and the male mowl.
- The yowl, howl or anger wail.
- The chatter (or chitter).

Murmur sounds

Produced with a closed mouth:

- The purr.
- Trills and chirrups.

Table 3.1. Methods of communication.

Method of communication	Advantages	Disadvantages
Visual signals (body language)	Clear and explicit Can be started, stopped or altered quickly	Intended recipient needs to be in close enough range to see the signaller clearly
Auditory signals (vocalizations)	The signaller does not have to be seen by the intended recipient Sounds can be effective over a long distance Can be started, stopped or altered quickly	The sound may be heard and exploited by an unintended receiver, e.g. an enemy or predator Sounds may become distorted by distance or overpowered by other sounds
Olfactory (scent) signals	Long lasting – information can be deposited by the signaller and read by the recipient when the signaller is no longer in the area. Ideal for solitary species, or animals at risk of predation	Information cannot be altered or retracted Contamination by microorganisms may alter the intended information

Strained intensity sounds

Produced with an open mouth. These are sounds that are most often used in offensive or defensive interactions, such as:

- Growl.
- Yowl.
- Shriek.
- Hiss.
- Spit.

The meow

The meow is the vocalization most commonly associated with the domestic cat; however, it is not a single distinct sound but a combination and selection of different sounds that can vary greatly between individuals and different breeds. It can also vary depending on the context and message that the cat intends to convey.

The meow is rarely used by feral or wildcats but is commonly used by pet cats in human-directed communication. It is therefore thought to be a product of domestication and socialization with people. It is certainly different to comparative vocalizations produced by wild or feral cats. A pet cat's meow tends to be shorter and higher pitched than similar sounds produced by feral cats (Nicastro, 2004; Yeon et al., 2011). Nicastro (2004) also found that the sounds produced by pet cats are more pleasing to the human ear than similar vocalizations produced by their wildcat relatives.

When directed towards people, a meow is used most often as a greeting and as a form of attention or food soliciting. Research has shown that most people can identify the meaning of a dog's bark, even if they have had very little experience with dogs (Pongrácz *et al.*, 2005). But the same is not true of the cat's meow. Cat owners can often distinguish the intention of their own cat's meows but even individuals with extensive experience with cats appear to be less able to identify the meaning of meows made by unfamiliar cats (Nicastro and Owren, 2003; Ellis *et al.*, 2015a). Meows therefore seem to be specific, not only to the context in which they are performed, but also to the individual cat. This could be because cats are able to learn, with repeated interaction and owner reinforcement, which sounds are most likely to be recognized and produce the desired reactions from their human caregivers.

A variation is the 'silent meow'. The cat opens and closes its mouth in exactly the same way as when producing a meow sound, and when directed towards people it is usually in the same circumstances as a normal meow but no sound is produced, at least not one that we can hear. Another difference is that the silent meow has been noted to be a part of cat-to-cat communication between feral and farm cats (Bradshaw *et al.*, 2012). It is possible that a sound may be produced but at too high a frequency for us to hear or it could be a part of visual signalling that we don't, as yet, fully understand.

The purr

The purr is produced by the contraction of the laryngeal muscles causing closure of the glottis. As pressure builds it forces the glottis open and causes the separation of the vocal folds, resulting in the purr. Unlike other vocalizations it is produced during both inhalation and expiration and can occur alongside other vocal sounds (Remmers and Gautier, 1972; Bradshaw *et al.*, 2012).

Like the meow, the purr is a sound that pet cats commonly direct towards humans, but unlike the meow it is also used in a wide range of cat-to-cat communication. Purring seems to be more than a means of communication because it also occurs when there are no other individuals, cat or human, around for the cat to communicate with (Kiley-Worthington, 1984, cited in Bradshaw *et al.*, 2012).

It has been found that pet cats produce two types of purr:

- The 'unsolicitation' purr is performed when the cat appears to be relaxed and content, either alone or in relaxed social contact.
- The 'solicitation' purr is performed when the cat is anticipating or soliciting food or attention. This purr also contains a higher pitched 'cry' and is reported as sounding more 'urgent' and less pleasant to human ears. Like the meow, this seems to be another form of specifically human-directed care-soliciting vocalization (Bradshaw and Cameron-Beaumont, 2000; McComb *et al.*, 2009).

Domestic cats are also known to purr when in extreme pain and even when dying. One hypothesis is that the low frequency vibrations caused by purring have a 'healing' effect (Von Muggenthaler, 2006). Other suggestions are that purring in such circumstances are a form of care or comfort soliciting or even possibly a form of self-reassurance (Bradshaw, 1992; Bradshaw *et al.*, 2012) but there is no scientific evidence as yet to support any of these theories, so the reason why cats do this remains unclear.

Defensive and antagonistic sounds

There are five sounds associated with defensive or antagonistic encounters. These are the strained intensity sounds: hiss, spit, growl, yowl and shriek. The hiss and spit are sounds of short duration, probably designed to initially deter a perceived threat such as potential predator or adversary. The growl and yowl, or caterwaul, are sounds of fairly long duration. Unlike the meow these are distinct sounds, although they are often used in combination with each other.

Generally speaking, in the animal world, the larger the animal the lower pitched its vocalizations. The growl is a low-pitched sound and one purpose of the growl might be to 'trick' an opponent or predator that the growling animal is larger than it actually is. The feline yowl, however, tends to be higher pitched than the growl.

If a perceived threat or opponent does not withdraw, growling and yowling can continue for some time and, possibly owing to tightness in the throat as a result of the cat's overall state of tension, the cat might start to drool and occasionally stop vocalizing to swallow. If the threat remains or increases, for example by moving closer, the cat may emit a sudden loud shriek. The purpose of this may be to startle the perceived threat or opponent and allow the cat a brief opportunity to escape or attack (Brown and Bradshaw, 2014). Sudden acute pain can also produce a shriek.

'Chattering'

Cats can sometimes be heard making a 'chattering' or 'chittering' sound when stalking prey or, more often, when potential prey can be seen but is unattainable, for example when viewed through a window. A popular idea is that the cat is attempting to mimic the sounds made by the prey, with the intention of attracting them. The margay (*Leopardus wiedii*), a South American wildcat, has been observed to seemingly mimic one of its major prey species, the pied tamarin (*Saguinus bicolor*). Although this does not seem to help with capture, it does appear to have the effect of drawing the prey closer to the predator and thereby reduces the energy expenditure of a pursuit (de Oliveira Calleia *et al.*, 2009). There seems to be no evidence, however, of any other cat species doing the same.

Other suggestions are that the chattering might be a displacement activity owing to frustration, or excitement and anticipation (McFarland, 1985, cited in Bradshaw *et al.*, 2012); however, the behaviour is not seen in other contexts where cats experience frustration or excitement. So until sufficient study is carried out and validated evidence becomes available, the purpose or cause of this form of vocalization is as yet unknown.

Sexual vocalizations

Entire (unneutered) males and females produce particular vowel-sound vocalizations during the breeding season. Mowl sounds, described by Shimizu (2001) as 'rutting' calls resembling the sound of a crying baby, are produced by males throughout the breeding season and a distinctive call is almost continually produced by females during the receptive period of the oestrus cycle. The most likely purpose of these vocalizations is to advertise strength and fitness to potential rivals and both fitness and availability to potential mates.

Vocalizations of kittens and nursing females

Kittens

During the first few days of a kitten's life its vocalizations are limited to a defensive spit and a distress call, which is emitted whenever the kitten is cold, hungry, isolated or trapped.

The occasions when the distress call is used, and its intensity and frequency, change as the kitten matures. For example, as the kitten's ability to regulate its own body temperature increases, its use of the call when cold decreases. Unwelcome restraint or being 'trapped' produces a high rate of distress calling up to around 6 weeks of age, whereas use of the distress call to indicate isolation tends to increase over the first 3 weeks before gradually decreasing. This could be due to the kitten becoming more aware of its mother and littermates during the first few weeks after birth, and therefore experiencing greater distress and anxiety when separated from them, before developing sufficient confidence and independence to be able to cope with separation. But it has also been found that the intensity of the distress call is greater the further away the kitten is from the nest (Mermet *et al.*, 2008). The increased use of the call might therefore also be associated with the kitten's increased physical ability, allowing it to stray further from the nest.

Overall use of the distress call starts to decrease at around 2–3 months of age and has usually stopped completely by the time the kitten is 5 months old (Bradshaw *et al.*, 2012).

Purring starts within a few days of birth and occurs mainly whilst suckling, which is possibly a way of signalling to the mother that the kitten is getting sufficient milk and to encourage continued nursing.

Aggressive vocalizations, including the defensive hiss, and the human-directed meow develop later, usually during or soon after weaning (Bradshaw *et al.*, 2012).

Nursing females

Most queens call to their kittens if they move away from the nest and produce a 'chirrup' or 'brrp' sound when close to or approaching their kittens. Szenczi *et al.* (2016) found that kittens respond more often and more positively to these sounds made by their own mother than to the same sounds made by another nursing queen, indicating that kittens are able to recognize their own mother's voice.

Both mothers and kittens purr during nursing and the mother cat will also often purr whilst grooming her kittens. Some mothers purr continuously whilst they are with their kittens (Lawrence, 1980; Deag *et al.*, 2000).

Visual Signalling

Cats use a wide range of facial expressions and body language but their visual signalling repertoire is nowhere near as varied or as complex as that of dogs. This is due in part to the musculature and resulting physical limitations of feline facial movements compared to that of dogs. But it is also due to specific differences between canine and feline ancestry and social structure influencing the need for complex visual signalling.

One of the major differences is that dogs, and the wolves they are descended from, are highly social animals and affiliative hunters. This means that they often need to share resources, especially food, so it is necessary for them to be able to signal appeasement and deference to others to avoid or lessen the likelihood of physical conflict (Bradshaw *et al.*, 2009; Hedges, 2014).

Cats are descended from a species that is predominantly solitary in nature, and although domestic cats are capable of group living, they are still solitary predators. They have not therefore developed, nor have the need for, intricate appeasement signalling. Conflict between cats is avoided by distancing, and aggressive or competitive encounters deflected with defensive measures rather than deference or appeasement.

Most feline visual signalling, especially between cats, can be defined as either distance-increasing – aimed to increase or maintain a distance from another individual – or distance-reducing – directed towards other individuals with which the cat wants to interact.

The following describes emotional states and individual elements of visual signalling. Care must be taken, however, not to attempt interpretation without also being aware of factors such as the situation that the cat is in at the time, plus concurrent behaviour and vocalizations. Looking at no more than a 'snap-shot' of a body posture or facial expression can easily lead to a misinterpretation of the animal's emotional state or signalling intent. It is also wise to be aware that a cat's mood, physical intent and associated signalling can change very quickly.

The face

The ears

A cat's ears are ideal for visual signaling: not only are they large and therefore highly visible, but also the musculature of the cat's pinnae (outer ears) allows for a wide range of movement that can be rapidly altered. The resting position of the ears is forward, although the ears also move frequently and independently to detect sounds. The following ear positions can be indicative of emotional states:

- Ears flattened sideways can be an indicator of fear. The more frightened the cat the more the flattened the ears. (Fig. 3.1).
- Frustration can be demonstrated by the ears being rotated backwards (Fig. 3.2). Finka *et al.* (2014) found that in cases of frustration or anger there appears to be greater rotation of the right ear.

The eyes

Dilated pupils, other than as reaction to reduced light, can be a sign of increased arousal rather than a specific mood state. So other visual signals and the cat's current situation should always be considered in order to identify the type and cause of arousal.

Fig. 3.1. Ears flattened to the side of the head can be a sign of fear. The more frightened the cat the flatter the ears. Notice also the dilated pupils, a sign of increased arousal.

Fig. 3.2. Frustration or irritation may be demonstrated by backwards rotation of the ears.

Rapid blinking can be a sign of fear and a way of avoiding direct eye contact in potential conflict situations. A slow blink that involves direct eye contact can, however, be a friendly gesture and if accompanied by relaxed body posture and purring can also signal contentment and relaxation (Fig. 3.3).

Other facial 'expressions' or movements

The tongue flick (or nose lick) can be an indicator of uncertainty or emotional conflict. It is unlike normal lip licking in that the tongue does not wipe around the mouth but 'flicks' up towards the tip of the nose (Fig. 3.4).

The positioning of the large mystacial vibrissae (facial whiskers) can also provide some insight into the cat's mood. When active and alert, for example during hunting and play, the whiskers spread out and point slightly forward and the use of the facial muscles to achieve this also gives the cat a slight 'puffy cheek' appearance.

Fig. 3.3. A slow eye-blink, or half-closed eyes can be a signal of contentment and relaxation.

Fig. 3.4. A tongue flick can be a sign of emotional conflict.

The whiskers are relaxed or might also be held slightly forward during positive encounters but are more likely to be held back against the face when fearful or at times of physical conflict.

The tail

A vertical raised tail, known as a 'tail-up' signal, is a distance-decreasing signal. It is usually a friendly greeting that often precedes allorubbing and sniffing (Fig. 3.5). Cameron-Beaumont (1997) found that cats would unhesitatingly approach a cat-shaped silhouette if the silhouette had a raised tail but were much less likely to approach a similar silhouette with a lowered or half-lowered tail.

Twitching the end of the tail is indicative of increased 'interest' and arousal rather than any specific mood state. This may be seen during play, when hunting, as well as when feeling frustrated, cautious or mildly irritated. More definite tail movements, however, such as swishing, slapping the tail on the ground while lying or sitting down, are a more definite sign of annoyance, frustration or feeling threatened and can often precede or accompany aggressive behaviour.

The 'bottle-brush' tail is part of a whole-body signal in response to a perceived threat and is a result of the hairs standing on end (piloerection) (Fig. 3.6).

Fig. 3.5. 'Tail up', a distance-decreasing signal that is often used as a friendly greeting.

Whole body signals

Distance increasing

Along with piloerection along the back and tail, arching the back to appear bigger and taller can be a direct response to a sudden severe threat such as a potential predator (Fig. 3.6). The tail may be positioned slightly to one side, towards the perceived threat, and is usually held down, although it might also be slightly raised at the base.

Although one defensive reaction is to appear bigger, another feline response to threat and potential danger is to attempt to appear much smaller by lowering the body towards the ground as much as possible, with the head lower than the body. If the cat has the opportunity, it will also attempt to hide (Fig. 3.7).

Encounters between rival cats can promote offensive and defensive body postures (Fig. 3.8):

- **Offensive body postures:** In antagonistic encounters with other cats it can be an advantage to appear confident and unafraid. To achieve this, the body carriage is generally raised and forward. The hairs on the body and tail may be raised (pilo-erection) to make the cat appear larger. The tail is held down and close to the body. The ears are forward and the eyes are open with intentional direct eye contact with the other cat.
- **Defensive body postures:** A defensive or fearful cat in an encounter is more likely to have a lowered body posture that is held back away from the challenger. The hairs on the body and tail may also be raised. The tail is held down and tight to the body. The ears are more likely to be flattened and/or rotated backwards. The other cat is watched closely but prolonged direct eye contact may be avoided by blinking, partial closure of the eyes and/or by occasionally looking slightly off to one side.

Fig. 3.6. Piloerection and arching the back makes the cat appear bigger than it actually is, and can be a response to a sudden severe threat such as a potential predator.

Fig. 3.7. A fearful cat may attempt to make itself appear smaller by lowered body posture with the head lower than the body. If it has the opportunity it will also attempt to hide.

Distance decreasing

Rolling, while stretching and opening the claws (known as a 'social roll'), may be used as an invitation to play and can be directed towards other cats, people, or even other animals that the cat has a friendly relationship with (Fig. 3.9). It has been described by some authors (Feldman, 1994a) as a sign of passive submission. But as cats have no need for deference signals and a social roll is more likely to occur when there is no threat from the other individual, this hypothesis seems unlikely.

Even though the social roll can be regarded as a friendly signal and an indication that a cat feels unthreatened, it should never be interpreted as a request by the cat to have its 'tummy rubbed' as the cat is already likely to be in an aroused (playful) state and doing so could result in being bitten and/or scratched. This could also be considered as a physical threat by the cat, provoking an aggressive defensive reaction.

Fig. 3.8. An aggressive encounter between rival cats; note the offensive and defensive body postures.

Fig. 3.9. A 'social roll' is a friendly signal, but it should never be interpreted as a request by the cat to have its 'tummy rubbed'.

Resting positions

The position in which a cat rests can indicate how tense or relaxed it is. Keeping the feet in contact with the ground allows the cat to 'spring' into action quickly. Resting in a crouched position with all four feet in contact with the ground can therefore be a sign that the cat is tense and 'ready for action', whereas a semi-tense cat may lie 'half-side' with just the front feet in contact with the ground. More relaxed resting positions are curled and lying completely on the side without any of the feet being in full contact with the ground. Lying or sleeping on the back is a less common resting position but it is one that cats can adopt when totally relaxed.

Tactile Communication

Tactile communication between cats includes sniffing, allorubbing (rubbing against each other) and allogrooming (grooming each other; allo – from the Greek *allos* meaning 'other'). When these behaviours occur mutually it can be a sign of a positive relationship between two or more cats. Attempts at friendly interactions are not always reciprocated, however, and might even be rebuffed with an aggressive response. Similar behaviours are also frequently directed towards humans and other animals, such as other household pets with which the cat has a good relationship. Some of these behaviours, such as rubbing and tail wrapping, might also be associated with olfactory communication because they allow the transfer of scent between individuals.

Sniffing

Nose touching (or simultaneous nose sniffing) is an affiliative behaviour that often, but not always, follows a mutual raised tail approach (Fig. 3.10). Observations of feral cat colonies have shown that this behaviour is more likely to be initiated by males than females (Cafazzo and Natoli, 2009). Investigatory sniffing of the rear or other areas of the body is more likely to follow a lowered tail approach so might be less likely to be directed towards a socially bonded or preferred individual (Brown, 1993).

Allorubbing

Mutual rubbing of heads, cheeks and flanks also occurs between socially bonded individuals and almost always follows a raised tail approach by the initiator (Brown,

Fig. 3.10. Nose touch, an affiliative greeting behaviour.

Fig. 3.11. Allorubbing occurs between socially bonded individuals and often follows a mutual tail-up approach.

1993) (Fig. 3.11). If both cats approach each other with a raised tail then both cats are more likely to rub against each other, but if the raised tail approach is unilateral the recipient might not respond at all or might only rub briefly after being rubbed on by the initiator. From studies of feral and farm cat colonies it appears that, unlike nose touching, allorubbing is more likely to be initiated by females than males and by younger group members towards older animals (Cafazzo and Natoli, 2009; Bradshaw *et al.*, 2012).

It has been speculated that rubbing between cats serves to maintain a 'colony odour' by exchanging scents (Bradshaw *et al.*, 2012). Another theory is that this could be an extension of a food solicitation behaviour that kittens direct towards their mother (Cafazzo and Natoli, 2009). This is more likely to be the case when the behaviour is directed towards human care givers at feeding time.

Tail wrap

This usually occurs between two socially bonded cats or might also be directed towards another animal or person considered to be friendly. The tail is directed so that it wraps around or against the other individual. When two cats are involved the tails usually wrap around each other. As with allorubbing the behaviour is more likely to follow a raised tail approach by the initiator and is more likely to be reciprocated if the recipient also has a raised tail.

Allogrooming

Grooming of another cat is usually directed towards the head and neck area and a cat may solicit grooming from another by approaching with a flexed neck to expose the underside of the neck and the side of the head (Crowell-Davis *et al.*, 2004). Except when performed by a mother cat to her kittens, allogrooming has a social rather than a cleaning function, although it is possible that the behaviour might have evolved from the mother–kitten relationship into a social affiliation behaviour. Allogrooming may also be another way of exchanging scents between individuals and maintaining a cohesive group scent. However, grooming of each other is not always an affiliative behaviour. The cat being groomed does not always appear to enjoy it and aggressive behaviour sometimes occurs during or after grooming with the groomer being the one more likely to act offensively (van den Bos, 1998).

Pet cats also learn various ways of soliciting grooming (stroking) from human owners including purring, head-butting, and tapping or pawing with a forefoot. Also, as when being groomed by another cat, most cats prefer to be stroked on or around the head, rather than on other parts of the body (Ellis *et al.*, 2015b). Pet cats will also occasionally lick people that they consider to be friendly or that they have a close attachment to, which might also be an attempt at mutual grooming.

Kneading

Purring and kneading, i.e. rhythmically pushing against an object with the front feet whilst extending and retracting the claws, is a behaviour of kittens that stimulates maternal milk flow. In adult cats it may be performed on soft material, e.g. bedding, other cats they are socially bonded with and towards human caregivers. Unlike rubbing, it does not seem to occur towards people so much when the cat is expecting food so, although it is a form of food soliciting in kittens, the purpose of the behaviour in adult cats is unclear. It may simply be a neotenic behaviour associated with pleasurable situations.

Olfactory Communication

Definitions

- **Semiochemical** (from the Greek *semeion*): a chemical substance used by plants and/or animals to communicate with or influence the physiology or behaviour of others.
- **Alleochemical**: the term used when the intended recipient or recipients of the semiochemical are members of a different species.
- **Pheromone**: the term used to define semiochemicals used in intraspecific communication, i.e between members of the same species, from the Greek *pherin* (to carry) and *hormon* (to stimulate) (Pageat and Gaultier, 2003). It can be argued, however, that not all intraspecific semiochemicals carry specific 'messages'. For example, the 'colony odour' achieved by feline allorubbing might be a socially cohesive scent rather than one conveying particular signals. Therefore, some authors prefer to refer to these as 'social odours'.

Many mammals, including the cat, use pheromones that can be 'read' by another individual of the same species via both normal olfaction (sense of smell) and the vomeronasal (Jacobson's) organ (Tirindelli *et al.*, 2009) (see Chapter 2). Pheromones are used in sexual communication but this is not their only function. They are used by cats via urine, faeces and secretions from skin glands in a variety of ways.

The skin glands

The skin glands associated with olfactory communication in the cat are found mainly around the head. These are:

- The submandibular gland under the chin.
- The perioral glands at the corners of the mouth.
- The temporal glands on either side of the forehead.
- The cheek glands.

The skin glands found on other parts of the body are:

- The interdigital glands on the feet.
- The supracaudal gland at the base of the tail.
- Unnamed groups of sebaceous glands along the tail.

All of these are sebaceous glands or groups of sebaceous glands. Most appear to produce very little or no obvious secretion except for the supracaudal gland, which is most evident in adult entire males and can sometimes over-secrete, resulting in the skin condition commonly known as 'stud tail' (supracaudal gland hyperplasia). Testosterone also influences the development of the cheek glands and associated sebum production in entire tomcats (Zielonka *et al.*, 1994).

Five different pheromones, each a different chemical compound but all containing varying amounts of fatty acids, have been isolated from the facial glands and what appears to be the function of three of them have been identified.

- F1 and F5: the functions of these have yet to be identified.
- F2: identified as being associated with sexual marking by entire males.
- F3: identified as being associated with facial marking of familiar and safe areas.
- F4: identified as the scent exchanged during allorubbing and allomarking (rubbing on the same inanimate object) by social group members (Pageat and Gaultier, 2003; Mills *et al.*, 2013).

Scent from the facial glands is deposited on to inanimate objects by a behaviour known as 'bunting' (Fig. 3.12). As well as scent-marking, it is possible that this activity might also be a part of visual communication.

A few of these facial pheromones have been synthetically reproduced and are commercially available. These may be used, alongside appropriate behavioural advice, to help prevent or treat behaviour issues (see Advising Clients in Chapter 11b).

Scratching

Cats scratch surfaces using the claws on the forefeet and often have preferred scratching areas where the behaviour is regularly repeated. They may demonstrate an individual preference for using either a vertical or horizontal surface or may scratch equally on both. One reason for scratching is to 'condition' the claws by helping to

Fig. 3.12. Scent is deposited onto inanimate objects by a behaviour known as 'bunting'.

remove the dead outer sheath. But scratching is also performed as a means of scent-marking. Scent is deposited during scratching from the interdigital glands between the toes and from glands in the large foot pad, known as the plantar pad. Scratching also leaves a visual mark that might serve the purpose of attracting other cats towards the olfactory mark.

The exact purpose of scratch marking is unclear. It is generally thought to be a form of territory marking because it is more likely to take place in familiar rather than unfamiliar areas (Feldman, 1994a, b). However, cats do not appear to avoid scratch marks made by other cats so it is unlikely to be a territorial warning. Mengoli *et al.* (2013) found that entire cats have an increased tendency to scratch mark in comparison to neutered individuals, and observations of feral cats have shown that the behaviour is more likely to occur in the presence of other cats (Turner, 1988, cited in Bradshaw *et al.*, 2012). It is possible that there might be an element of sexual 'advertising'. But neutered cats also scratch and with varying enthusiasm. Owners often report that their cats scratch furniture or scratch posts when they return home or when they greet the cat first thing in the morning, and Schär (1986) cited in Mertens and Schär (1988) suggested that scratching might be a sign of excitement during interactions with people. However, scratching by pet cats occurs both indoors and outdoors, in the company of people, other cats and when alone, which indicates that the message the cat is leaving may be as much for itself as it is for others and one theory is that it might provide an indicator for the cat of where he can relax and where he may need to be vigilant (Casey, 2009).

Urine

Urine marking or 'spraying' is probably the most recognized of the feline methods of scent signalling. When spraying, a cat adopts a particular standing position, with the tail held vertically and quivering, with the hind feet treading alternately. In this position a stream of urine is directed backwards, usually onto a vertical surface (Fig. 3.13). The amount of urine passed can vary and may depend on how full the cat's bladder is at the time (Neilson, 2009).

After normal squat urination cats will attempt to bury or cover the urine, but there is usually no attempt to cover spray marks and the urine is deposited at cat 'nose height' so that it can be clearly evident to other cats. Cats show a lot more interest in spray marks than in urine eliminated via squatting and investigation of a spray mark can often elicit a 'flehmen' response (see Chapter 2), activating the vomeronasal organ (Bradshaw *et al.*, 2012).

Urine spraying is most frequently and commonly performed by entire males and more often when in the vicinity of a female in oestrus. Entire, sexually receptive females also urine mark (Bradshaw *et al.*, 2012). Therefore, urine spraying is most likely to be a means of transferring information about fitness and sexual status (see Chapter 4). But all adult cats, regardless of sex and neuter status, can and often do leave scent marks in this manner so sexual marking cannot be the only reason for the behaviour.

Indoor urine spraying by neutered cats seems to occur when the cat is anxious and/or stressed (Amat *et al.*, 2015). It is also more likely to be performed in the area where the cat has previously felt insecure or at threat. There are many possible causes of feline stress and anxiety but when associated with urine spraying the cause is most often real or

Fig. 3.13. Urine marking (spraying). (Source: Commander-pirx at German Wikipedia, under a CC BY-SA 3.0 license.)

perceived competition with other cats, either those living in the same household or neighbouring cats that invade or are perceived as potential invaders to the cat's core territory.

Spraying could therefore be a territorial behaviour but it is unclear what message is being conveyed. Hypotheses include the following:

- It might be a simple warning to other cats. Cats do not, however, appear to avoid spray marks made by other cats; in fact, they will often spend a long time investigating them (Bradshaw and Cameron-Beaumont, 2000). It is therefore unlikely that spraying is performed solely as a means of contact avoidance, or as a warning to an 'intruder' to leave the territory.
- The scent might be deposited to allow an intruder to recognize the territory owner by comparison of the scent deposited and the scent of the cat that left the scent mark. This could only work, however, if the two cats had already met.
- The urine mark might contain information about the depositor's health and competitive ability, allowing the invader to consider its chances of winning a confrontation and the potential risks involved.
- Because the chemicals in the urine degrade over time, the strength of the scent might give the investigator some idea of how long ago the scent mark was produced so when the depositor was last in the area.
- Leaving a scent mark might even be a coping strategy for the depositor, either by increasing his scent profile or by allowing increased predictability and control by providing himself with information of where extra vigilance may be required (Bowen and Heath, 2005; Bradshaw *et al.*, 2012).

Faeces

Although many carnivores use faeces as a part of olfactory communication, there is no definite evidence that this is the case with domestic cats. Nakabayashi *et al.* (2012) found that cats spend longer sniffing the faeces of unfamiliar cats compared to the time spent sniffing their own or that of a familiar cat, suggesting that some olfactory information might be conveyed in faeces. But further studies are needed. It has also been observed that feral and farm cats are more likely to bury their faeces close to the core territory and leave those deposited in hunting areas or at the boundary of the territory exposed, suggesting the possible use of faeces as a territory marker (Feldman, 1994b; Bradshaw and Cameron-Beaumont, 2000). But another explanation for this behaviour is that there is simply greater incentive to bury faeces closer to the core territory for hygienic purposes and to avoid attracting potential predators.

References

Amat, M., Camps, T. and Mantecu, X. (2015) Stress in owned cats, behavioural changes and welfare applications. *Journal of Feline Medicine and Surgery* 18, 577–586.
Bowen, J. and Heath, S. (2005) *Behaviour Problems in Small Animals. Practical Advice for the Veterinary Team*. Saunders Ltd, Elsevier, Philadelphia, Pennsylvania, USA.
Bradshaw, J.W.S. (1992) *The Behaviour of the Domestic Cat*. CAB International, Wallingford, UK.
Bradshaw, J.W.S. and Cameron-Beaumont, C. (2000) The signalling repertoire of the domestic cat and its undomesticated relatives. In: Turner, D.C. and Bateson, P. (eds) *The Domestic*

Cat: The Biology of its Behaviour, 2nd edn. Cambridge University Press, Cambridge, UK. pp. 67–93.

Bradshaw, J.W.S., Blackwell, E.J. and Casey, R.A. (2009) Dominance in domestic dogs – useful construct or bad habit? *Journal of Veterinary Behavior* 4, 135–144.

Bradshaw, J.W.S., Casey, R.A. and Brown, S.L. (2012) *The Behaviour of the Domestic Cat*, 2nd edn. CAB International, Wallingford, UK.

Brown, S.L. (1993) The social behaviour of neutered domestic cats *(Felis catus)*. PhD thesis, University of Southampton, Southampton, UK.

Brown, S.L. and Bradshaw J.W.S. (2014) Communication in the domestic cat: within- and between-species. In: Turner, D.C. and Bateson, P. (eds) *The Domestic Cat: The Biology of its Behaviour*, 3rd edn. Cambridge University Press, Cambridge, UK pp. 37–59.

Cafazzo, S. and Natoli, E. (2009) The social function of tail up in the domestic cat *(Felis silvestris catus)*. *Behavioural Processes* 80, 60–66.

Cameron-Beaumont, C.L. (1997) Visual and tactile communication in the domestic cat *(Felis silvestris catus)* and undomesticated small felids. Ph.D. thesis, University of Southampton, Southampton, UK.

Casey, R.C. (2009) Management problems in cats. In: Horwitz, D.F. and Mills, D.S. (eds) *BSAVA Manual of Canine and Feline Behavioural Medicine*, 2nd edn. BSAVA Gloucester, UK.

Crowell-Davis, S.L., Curtis, T.M. and Knowles, J. (2004) Social organization in the cat: a modern understanding. *Journal of Feline Medicine and Surgery* 6, p. 19–28.

Deag, J.M., Manning, A. and Lawrence, C.E. (2000) Factors influencing the mother–kitten relationship. In: Turner, D.C. and Bateson, P. (eds) *The Domestic Cat: The Biology of its Behaviour*, 2nd edn. Cambridge University Press, Cambridge, UK, pp. 23–45.

de Oliveira Calleia, F., Rohe, F. and Gordo M. (2009) Hunting strategy of the margay (*Leopardus wiedii*) to attract the wild pied tamarin (*Saguinus bicolor*). *Neotropical Primates* 16, 32–34.

Ellis, S.L.H., Swindell, V. and Burman, O.H.P. (2015a) Human classification of context-related vocalizations emitted by familiar and unfamiliar domestic cats: An Exploratory Study. *Anthrozoös* 28, 625–634.

Ellis, S.L.H., Thompson, H., Guijarro, C. and Zulch, E. (2015b) The influence of body region, handler familiarity and order of region handled on the domestic cat's response to being stroked. *Applied Animal Behaviour Science* 173, 60–67.

Feldman, H.N. (1994a) Domestic cats and passive submission. *Animal Behaviour* 47, 457–459.

Feldman, H.N. (1994b) Methods of scent marking in the domestic cat. *Canadian Journal of Zoology* 72, 1093–1099.

Finka, L., Ellis, S.L.H., Wilkinson, A. and Mills, D. (2014) The development of an emotional ethogram for *Felis sivestris* focused on FEAR and RAGE. *Journal of Veterinary Behavior* 9(6), e5.

Hedges, S. (2014) *Practical Canine Behaviour for Veterinary Nurses and Technicians*. CAB International, Wallingford, UK.

Kiley-Worthington, M. (1984) Animal language? Vocal communication of some ungulates, canids and felids. *Acta Zoologica Fennica* 171, pp. 83–88. Cited in: Bradshaw, J.W.S., Casey, R.A. and Brown, S.L. (2012) *The Behaviour of the Domestic Cat,* 2nd edn. CAB International, Wallingford, UK.

Lawrence, C.E. (1980) Individual differences in the mother-kitten relationship in the domestic cat *(Felis catus)*. PhD thesis, University of Edinburgh, Edinburgh, UK.

McComb, K., Taylor, A.M., Wilson, C. and Charlton, B.D. (2009) The cry embedded within the purr. *Current Biology* 19 R507–R508.

McFarland, D., (1985) *Animal Behaviour: Psychobiology, Ethology and Evolution*. Longman, Harlow, UK.

Mengoli, M, Mariti, C., Cozzi, A., Cestarollo, E., Lafont-Lecuelle, C., Pageat, P. and Gazzano, A. (2013) Scratching behaviour and its features: a questionnaire-based study in an Italian sample of domestic cats. *Journal of Feline Medicine and Surgery* 15, 886–892.

Mermet, N., Coureaud, G., McGrane, S. and Schaal, B. (2008) Odour-guided social behaviour in newborn and young cats: an analytical survey. *Chemoecology* 17(4), 187–199.

Mills, D., Dube, M.B. and Zulch, H. (2013) *Stress and Pheromonatherapy in Small Animal Clinical Behaviour*. John Wiley & Sons, Oxford, UK.

Nakabayashi, M., Yamaoka, R. and Nakashima, Y. (2012) Do faecal odours enable domestic cats (*Felis catus*) to distinguish familiarity of the donors? *Journal of Ethology* 30, 325–329.

Neilson, J.C. (2009) House soiling by cats. In: Horwitz, D.F. and Mills, D.S. (eds) *BSAVA Manual of Canine and Feline Behavioural Medicine*, 2nd edn. BSAVA, Gloucester, UK.

Nicastro, N. (2004) Perceptual and acoustic evidence for species-level differences in meow vocalisations by domestic cats (*Felis catus*) and African wild cats (*Felis sylvestris lybica*). *Journal of Comparative Psychology* 118, 287–296.

Nicastro, N. and Owren, M.J. (2003) Classification of domestic cat (Felis catus) vocalizations by naïve and experienced human listeners. *Journal of Comparative Psychology* 117, 44–52.

Pageat, P. and Gaultier, E. (2003) Current research in canine and feline pheromones. *Veterinary Clinics of North America. Small Animal Practice* 33, 187–211.

Pongrácz, P., Molnár, C., Miklósi, A. and Csányi V. (2005) Human listeners are able to classify dog (*Canis familiaris*) barks recorded in different situations. *Journal of Comparative Psychology* 119, 136.

Remmers, J.E. and Gautier, H. (1972) Neural and mechanical mechanisms of feline purring. *Respiration Physiology* 16, 351–361.

Schär, R. (1986) Einfluss von Artgenossen und Umgebung auf die Sozialstruktur von fünf Bauernkatzengruppen. Lizentiatsarbeit. Bern, Druckerei der Universität Bern. Cited in: Mertens, C. and Schär, R. (1988) Practical aspects of research on cats. In: Turner, D.C. and Bateson, P. (eds) *The Domestic Cat: The Biology of its Behaviour*. Cambridge University Press, Cambridge, UK.

Shimizu, M. (2001) Vocalizations of feral cats: sexual differences in the breeding season. *Mammal Study* 26, 85–92.

Szenczi, P., Bánszegi, O., Urrutia, A., Faragó, T. and Hudson, R. (2016) Mother–offspring recognition in the domestic cat: Kittens recognize their own mother's call. *Developmental Psychobiology* 58, 568–577.

Tirindelli, R., Dibattista, M., Pifferi, S. and Menini, A. (2009) From pheromones to behavior. *Physiological Reviews* 89, 921–956.

Turner, D.C. (1988) Cat behaviour and the human/cat relationship. *Animalis Familiaris* 3, 16–21, Cited in: Bradshaw, J.W.S., Casey, R.A. and Brown, S.L. (2012) *The Behaviour of the Domestic Cat,* 2nd edn. CAB International, Wallingford, UK.

Van den Bos, R. (1998) The function of allogrooming in domestic cats *(Felis silvestris catus)*: A study in a group of cats living in confinement. *Journal of Ethology* 14, 123–131.

Von Muggenthaler, E. (2006) The felid purr: A bio-mechanical healing mechanism. In: *Proceedings of the 12th International Conference on Low Frequency Noise and Vibration and its Control.* Bristol, UK.

Yeon, S.C., Kim, Y.K., Park, S.J., Lee, S.S., Lee, S,Y. et al. (2011) Differences between vocalizations evoked by social stimuli in feral cats and house cats. *Behavioural Processes* 87, 183–189.

Zielonka, T.M., Charpin, D., Berbis, P., Luciani, P., Casanova, D. and Vervloet, D. (1994) Effects of castration and testosterone on *Fel d 1* production by sebaceous glands of male cats. *Clinical and Experimental Allergy* 24, 1169–1173.

4 Social, Feeding and Predatory Behaviour

Social Behaviour

Group living among members of the cat family is rare. Other than lions, almost all members of the cat family lead solitary lives as adults, including the ancestor of the domestic cat, the African wildcat (*Felis silvestris lybica*) (Bradshaw, 2016a). The domestic cat, however, is highly flexible in its sociality, having the ability to live alone or with others, and can even develop close social bonds with conspecifics.

The ability to be social probably developed as part of early domestication (see Chapter 1), when otherwise solitary animals would have been drawn together towards human settlements by large numbers of rats and mice attracted there by human waste and food stores. Man-made structures would also have provided ideal protection, close to an ample food source, for females raising their kittens. The cats that were best able to tolerate being in close proximity to other cats would therefore have had increased hunting opportunities and better survival prospects, and as a result been more likely to pass their genes on to future generations. As human settlements grew in size and density, and the number of cats living near or within towns and villages increased, the ability to be tolerant and to live alongside other cats would have become even more important (Bradshaw, 2016b).

However, changes to species-specific behaviour is a very slow ongoing process and the domestic cat has not yet evolved to be fully social and there are limitations and constraints to this ability, so it should never be assumed that cats will instinctively get along with each other just because they are members of the same species. In fact, a major source of stress for cats is enforced close proximity and resource sharing with other cats, especially with those not considered to be part of the same social group.

The social behaviour of feral cats

Feral cats should not be confused with wildcats or domestic strays (ownerless cats that were once pets). By definition a feral cat is a domestic cat (*Felis silvestris catus*) that has never been socialized with people and is not solely reliant on a human caregiver for food, shelter or care.

Studies have shown that groupings of feral cats are most likely to occur where access to prey or other food resources are plentiful and predictable. Farmyards, refuse dumps or areas where there are frequent handouts from people are the most likely places to find feral cat colonies. In areas where food resources are sparse cats are more likely to be solitary (Corbett, 1979; Liberg *et al.*, 2000).

This raises the question as to whether aggregations of feral cats are evidence of social bonding or just gatherings of otherwise solitary animals around a plentiful food source; this is something that is often seen in the wild and among feral or free-ranging domestic animals where members of different species that would normally regard each other as prey or predator will eat or drink close to each other, especially if the resource is of high value or otherwise limited. But observations of friendly interactions between feral cats found in groups indicate that many are socially bonded and not just attracted to the same location by a food source (Kerby and Macdonald, 1988; Crowell-Davis *et al.*, 2004).

The availability of shelter and suitable nesting sites is another relevant factor that might explain why colonies of feral cats are more likely to exist in areas such as farms and industrial sites where barns, warehouses and disused buildings can provide ideal accommodation.

The size of a feral cat colony can vary greatly from fewer than ten to more than 50 members, although larger colonies are more likely to be made up of smaller social groups. These colonies consist mainly of related females and their offspring, including sexually immature and some young adult males. Other than when mating, contact with cats who are not part of the same group is avoided and members of the group are hostile to outsiders who may try to invade the territory or pose a threat to resources (Kerby and Macdonald, 1988; Bradshaw *et al.*, 2012).

What is the advantage of social grouping?

Because cats are self-reliant hunters, individual survival is not dependent on group co-operation. But being in a group, where there is a collective effort to defend resources, can be advantageous for a breeding female and her kittens. It has also been observed that group-living females help each other with the birth, care and feeding of each other's litters (Allaby and Crawford, 1982; Macdonald *et al.*, 1987), a behaviour that is also seen in owned breeding queens, especially related females. A disadvantage of living with others, however, can be an increased risk of disease. Group-living feral cats have been shown to have a far higher pathogen prevalence than household domestic cats (Macdonald *et al.*, 2000).

Mature males

Once they reach sexual maturity most males will leave their mother's home range to lead a more solitary existence. Sexually mature entire males will either avoid each other or fight whenever they meet, especially when there is a female in oestrus nearby (Dards, 1983). However, males that are littermates or have been raised within the same colony have been seen to form coalitions or 'brotherhoods', demonstrating amicable greeting behaviours, and even patiently 'waiting their turn' to alternately copulate with an in-oestrus female (Allaby and Crawford, 1982; Liberg and Sandell, 1988; Macdonald *et al.*, 2000; Crowell-Davis *et al.*, 2004). It is possible, however, that this only happens with juvenile males and that aggression is the more likely result of encounters between more mature males (Dards, 1983; Bradshaw *et al.*, 2012).

Home ranges

The 'home range' is the area over which a cat will regularly roam. This includes:

- **The territory**: the area that the cat will actively defend from other cats that are not a part of the same social group; and
- **The core territory**: situated within the territory. The area where the cat feels most secure. For pet cats this is usually within its owner's home.

An intact (unneutered) male's home range is on average around three and half times larger than that of a female (Bradshaw *et al.*, 2012). The size of the home range is influenced by what the cat considers to be of most importance, including the number of other cats in the area; for females this is overall cat density and for males it is the number of receptive females in the area. For both sexes, the availability of food and/or prey is also important (Liberg and Sandell, 1988). A male's home range can vary in size at different times of the year, being generally larger during the mating season but this is not the case for neutered cats, which show no seasonal difference in home range size. Neutering also seems to significantly alter the size of the home range for both sexes, being much smaller for neutered than sexually entire cats. All cats, regardless of sexual status, appear to roam further at night than during the day (Thomas *et al.*, 2014; Kitts-Morgan *et al.*, 2015).

The social behaviour of pet cats

The effects of neutering

The majority of owned cats in the UK over the age of 6–12 months are neutered (Murray *et al.*, 2009). Neutering can have a significant influence on a cat's behaviour, especially social behaviour with other cats.

Neutering is the removal of the reproductive organs to prevent sexual activity and unwanted pregnancies. It also prevents, alters or modifies other hormonally influenced behaviour. Neutering to prevent unwanted behaviour is most successful if performed at or before puberty.

Castration is the removal of the testes (testicles) of male cats. Castration can significantly reduce sexually motivated behaviour such as urine marking and fighting with other males. Castration increases the chances, but does not guarantee, that a male cat will be able to live alongside other cats.

Spaying is the removal of the uterus and ovaries of female cats. Spayed females do not have reproductive cycles (see Chapter 5) so do not attract males or display behaviours associated with being in oestrus.

Social groupings in multi-cat households

The number of cats kept in individual households can vary tremendously, as can the housing conditions and availability of space and resources. It should never be assumed that because pet cats live together they are, or will become, close companions, even if they have had a close feline companion previously. Unlike wild, stray or feral cats, house cats have no choice as to the number, sex or relatedness of other cats they are compelled to share their home with.

Most multi-cat households are made up of a combination of solitary individuals and small social groups, with often no more than two to three in each group, although this can vary. Even in a house with numerous cats, it is possible that none of them is socially bonded and that all are solitary individuals.

Social bonding can and does occur between pet cats, although it is more likely to exist between cats that have lived together from a very young age, especially if closely related (Bradshaw and Hall, 1999; Curtis *et al.*, 2003). It is certainly not unheard of for cats introduced when one or both are adult to become friends but the chances of this happening are far less than if they had grown up together.

Fighting and stress-related issues are more commonly seen in cats introduced as adults. In a study by Levine *et al.* (2005), 375 owners who had adopted an adult cat from a New York animal shelter were questioned. 49.6% reported initial fighting between the new cat and cat(s) already owned, and 35% reported that fighting was continuing 2–12 months later.

The fact that the remaining 65% were reported not to be fighting a year later does not necessarily indicate that these cats became close companions. Even between unneutered males, the frequency and intensity of fighting between the same individuals will reduce over time so it is very probable that the same will occur between neutered pets. It is also common for owners to misinterpret mild to moderate aggression as play. But even if cats are no longer fighting or showing discernible signs of conflict this is not sufficient to indicate social bonding. In many cases cats who do not regard each other as members of the same social group can learn to live together in a state of mutual tolerance or resigned acceptance.

The ability to tolerate the presence of other, previously unknown cats within the core territory can be highly dependent on factors such as previous experience, especially early socialization with other cats, and the individual's innate ability to cope with challenging situations, which in turn can be affected by things such as general health and exposure to stressors during development (see Chapter 5).

Cats living in the same household can also learn methods of conflict avoidance such as time-sharing access to resources including food, resting places, litter trays and even access to the owner. But environmental limitations, or an owner's mistaken belief that their cats are good friends, might impinge on their ability to avoid conflict, resulting in fighting, long-term stress and an increased likelihood that undesirable behaviours and/or stress-related health issues may develop.

Even when there is close social bonding between household cats a relationship might not always be permanent, because it can easily be damaged or break down completely if exposed to challenges such as health issues, external stressors and limited resources, resulting in competition between the cats.

Signs of social bonding

Signs of social bonding include:

- Seeking out each other's company.
- Vertically raised tail approach and greeting by sniffing/touching noses.
- Mutual grooming (allogrooming).
- Rubbing against each other (allorubbing).
- Resting and sleeping close together, usually touching, sometimes wrapped around each other (Fig. 4.1).

Fig. 4.1. Cats that are socially bonded are more likely to sleep or rest in close contact with each other. Those that are not of the same social group may share a preferred resting area, but maintain a distance from each other.

Play can also be a sign of social bonding and may be differentiated from fighting by the following:

- Being generally quiet, with very few or no vocalizations.
- Both cats show forward directed body posture for a lot of the time during play.
- Claws are retracted and bites inhibited.
- Following play, the cats return to bonding behaviour as described above.

Although one individual might consider another as a friend, the feeling might not be reciprocated so affiliative behaviours can sometimes be directed one way only.

Signs often misinterpreted as evidence of social bonding

FIGHTING. Fighting will occasionally occur between cats that otherwise have a friendly relationship, but if the cats are fighting more often than playing it is a good sign that they are not getting along. A common problem, however, is that it can sometimes be difficult to distinguish the difference between play fighting and real fighting. The following behaviours are more likely to be seen if the cats are fighting rather than playing:

- Vocalizations – especially growling, hissing or spitting.
- Head and body posture held back while defensive swipes are made with front feet.
- Ears rotated back and/or down to the sides.
- Dilated pupils.
- Claws extended, bites less inhibited.
- 'Swishing' tail.

EATING TOGETHER. It is often assumed that if cats eat alongside each other they have a good relationship. In many cases, however, the reason they eat together is because they have no other option, especially if food bowls are placed close to each other and there is no other equally nutritious food available. Time-sharing access to the food may not be an option because if one cat leaves the feeding area the other cat(s) might

Fig. 4.2. Eating together is not a sign of a good relationship. Cats can be forced to eat close together if there is no equally nutritious food available elsewhere, a situation that can be highly stressful for the cats and might increase rivalry and aggression.

eat his share of the food. It may therefore appear that cats are at ease eating together but in truth this situation can be highly stressful for them. Being forced to eat together like this can also increase their perception that food is limited and so further increase competition and rivalry. (Fig.4.2).

RESTING TOGETHER. Even cats who do not get along will often rest or sleep in the same area or on the same item of furniture, but their exact positioning can reveal that they are simply using the same preferred resting place rather than demonstrating any desire to be with each other. In such cases, body postures might not be completely relaxed and there can often be a discernible distance between them (Fig. 4.1).

Signs (other than fighting) of an antagonistic relationship

AVOIDANCE. The cats may spend most of their time in separate areas of the house or, if allowed outdoor access, spend more time away from home. If in the same room, they may make use of raised vertical spaces, e.g. tops of furniture etc., to avoid encounters.

WATCHING. Close or constant watching of one cat by another can be an indicator that the cat is feeling threatened by the presence of the other cat.

GUARDING ACCESS TO RESOURCES (BLOCKING). An increased sense of competition can result in guarding resources by blocking the other cat's access. This is usually achieved by the guarding cat positioning itself in a corridor or at the entrance to a room containing a valued resource such as food, litter tray, resting place or even access to the owner. Blocking

Fig. 4.3. One cat may intentionally block another cat's access to important resources such as food, resting places, or litter trays.

can often be sufficient to prevent the other cat attempting to gain access but, if not, it might be attacked as it attempts to get past (Fig. 4.3).

AMBUSHING. Ambush attacks are another behaviour that might be misinterpreted as play and can also be a part of resource guarding. These attacks often take place when and where the victim is most vulnerable, such as during or after using a litter tray, especially when exiting a covered tray, and when entering or exiting the house via a cat flap.

Dominance hierarchies

Dominance is a term that describes a position of power or advantage over others. An individual who is victorious in a conflict situation and/or achieves priority access to a resource may be described as dominant. But the term can only really be applied to the individual with regard to the immediate outcome of the conflict or competitive situation. Dominance or 'being dominant' cannot be used to describe an animal's temperament or behaviour.

A dominance hierarchy can develop, however, within a group or dyad when the dominance/subordinate relationship becomes stabilized or is maintained, sometimes by aggressive and/or deferential interactions. For some group-living species this can be part of their normal social structure.

The question is whether dominance hierarchies exist in feral cat colonies or within multi-cat households? Certainly where cats are placed in a situation of having to share limited living space and resources, competition between them is almost inevitable, sometimes resulting in more confident and assertive individuals gaining greater access to resources. But whether this can be regarded as evidence of a dominance hierarchy is debatable.

The domestic cat's primary ancestor *Felis s. lybica* is primarily a solitary and highly territorial animal, so it seems highly unlikely that dominance hierarchy is a part of the domestic cat's ancestral social structure. But social bonding and group living, believed to be by-products of domestication, are recognized behaviours of the domestic cat. Indeed, some early laboratory-based studies described cats as having dominance hierarchies on the basis of the observations that some cats can achieve priority access to food (Masserman and Siever, 1944; Cole and Shafer, 1966). However, later studies on feral cat groups found that, although episodes of aggression between group members can occur, there appears to be more evidence of coalition than status-related conflict (Macdonald *et al.*, 2000; Bradshaw, 2016a). Also, for dominance hierarchies to be maintained, subordinate group members need to be able to defer and demonstrate submission to higher-ranking individuals. Deferential signalling is very limited in the cat and aggressive or competitive encounters are more likely to be avoided or deflected with defensive rather than submissive behaviours (Bradshaw *et al.*, 2012). Also, as self-reliant predators, with each individual hunting to provide food only for him or herself, cats do not need the signalling repertoire necessary to avoid conflict over shared food resources. It therefore seems unlikely that dominance hierarchies are a normal part of the domestic cat's social structure.

Neighbourhood cats

Relationships between neighbouring cats are rarely amicable and fights are not uncommon, especially in areas of high feline population density. Neutering can reduce the propensity for fighting because this removes sexual competition but only to a point as space and access to hunting areas are also important resources to be defended. Cats will also feel the need to defend their core territory and the resources within it from potential feline invaders.

Fighting is not always inevitable, however, between neighbouring cats because they can also employ means of conflict avoidance such as time-sharing access to specific areas, hiding, or accessing high places from where they can observe rivals and physically avoid encounters.

Social behaviour with people

There is no doubt that close social relationships do develop between pet cats and their owners, although the strength and quality of the cat's relationship with the owner can depend greatly on factors such as individual temperament and early socialization with people (see Chapter 5) plus the general treatment, care and handling of the cat by the owner.

How do our pet cats view us?

It is unknown exactly how our pet cats regard us. There is a common belief that they view us as caregivers, as a kitten does its mother. This idea can be reinforced by the fact that some cats direct kneading and even suckling behaviours towards their owners, but not all pet cats do this and these behaviours can be just as frequently directed onto inanimate objects, especially soft, fur-like fabrics. Another idea is that because pet cats demonstrate the same affiliative behaviours towards people as they do towards other cats considered to be part of the same social group, they regard us as fellow social group members. But as there is no validated scientific evidence to support either hypothesis, the cat's view of its role in its relationship with us remains a mystery.

Cats often tend to favour the person that feeds them. Geering (1986, in Turner, 2000) found that although other interactions such as stroking and playing were necessary to help maintain a relationship, feeding was most important in establishing the relationship in the first place. A recent study by Shreve *et al.* (2017) suggested, however, that cats prefer human interaction to food, although the social interaction with people that most of the cats in the study preferred was playing with a wand type toy. When playing, a cat's focus of attention will be far greater towards the toy than towards the human who is making the toy move, so this could be seen more as an interaction with the toy than with the person. The other interactions offered were being stroked or being talked to, which most of the cats preferred less than playing with the toy. One specific food, tuna, was also preferred over being stroked or spoken to, although preferred less than interactions with the toy, indicating that in fact predatory-type play was actually the most preferred activity, followed by tasty food and then human interaction.

But it would be impossible to extrapolate the preferences of all cats from a single study because this can vary greatly between individuals and be influenced by multiple factors such as time of day, mood, hunger, and the strength and quality of the relationship with the person.

Attachment to owners

An attachment is an emotional bond between two or more individuals. Inarguably pet cats develop a bond and affection for their owners, although there is some debate and further research required as to the type or level of attachment that cats have to their human caregivers.

In relationships of secure attachment, an individual perceives another as a point of safety and security and can become highly distressed when separated from their attachment figure, a condition commonly known as 'separation anxiety'. Secure attachment is the bond that exists between the young of many species, including human children, and their mothers, and has been well documented to exist between pet dogs and their owners (Bowlby, 1988; Topál *et al.*, 1998; Prato-Previde *et al.*, 2003).

Some authors report that cats also exhibit secure attachment to their owners (Schwartz, 2002; Edwards *et al.*, 2007). But a more recent study questioned previous findings and revealed that although cats are affectionate towards their owners, they do not appear to demonstrate this particular form of attachment (Potter and Mills, 2015). In another, albeit small and unpublished study by Zak (2016), blood samples were taken from both dogs and cats before and after interacting with their owners, and tested for oxytocin. Oxytocin is a hormone and neurotransmitter that is often called 'the love hormone' because it plays an important role in emotional attachment. After interacting with their owners, the average increase in oxytocin in the dogs was

Fig. 4.4. It is unknown exactly how our pet cats regard their relationship with us. Some recent research suggests that they may consider us as 'good friends'.

57% but in the cats it was 12%, which is about the same level as seen in people when meeting friends. This difference in oxytocin levels between dogs and cats would seem to support the theory that dogs are more likely to develop a secure attachment to their owners, possibly seeing them as essential caregivers, whereas cats are more likely to regard their owners as good friends (Fig. 4.4).

When considering the differences between canine and feline social structure and ancestry, plus the fact that dogs exhibit a much higher level of neoteny than cats, it would seem very probable that cats, who are typically more autonomous in nature than dogs, are far less likely to develop a secure attachment towards their owners.

Cats have, however, been reported to express behaviours that could be attributed to separation anxiety (Schwartz, 2002). But further research is required because it is possible that there could be other reasons for these behaviours such as frustration, or fear and anxiety related to events, rather than anxiety due solely to separation from the owner.

Feeding Behaviour

Food preferences

Cats require a higher level of protein in their diet than many other mammals. As well as being a major source of energy and necessary for tissue growth and repair, cats have an increased need for certain amino acids, the molecule chains that form proteins, that they are unable to synthesize from other foods or conserve within their bodies in sufficient quantity. These amino acids – taurine, arginine, methionine and cysteine – can only be found in sufficient quantity in animal flesh. Deficiency can result in blindness, and heart, neurological and immune system abnormalities (Zoran, 2002). Cats are therefore obligatory carnivores; in other words, they need to eat meat to remain healthy and the foods that most cats choose to eat reflect this.

Cats have far fewer taste receptors (taste buds) than either dogs or humans, having around 475 in comparison to 1700 in dogs and 9000 in humans (Horwitz *et al.*, 2010). But they can detect the taste of specific amino acids and show a preference for those that we perceive as sweet, although they have no ability to detect sugar and are therefore unlikely to show any preference or aversion to otherwise sweet substances. Artificial sweeteners, such as saccharin, however, tend to be rejected, possibly because a cat will perceive them as bitter rather than sweet (Bartoshuk *et al.*, 1975). Cats have a particular sensitivity to and dislike of bitter tastes. A preference for sweet rather than bitter amino acids might also help them to identify and prefer fresh meat, which is higher in the amino acids they require than meat that has started to decompose.

Olfaction plays a major part in food acceptance; cats will often reject food if they cannot smell it. This explains why most prefer food that is at room or blood temperature rather than cold food straight out of the fridge, which will have a much-reduced odour. For feral and wildcats, this might also be a way of identifying freshly killed prey over carrion.

Some cats will show an increased preference for a new food when offered and may even reject their previous food. This could be a natural behaviour to encourage dietary diversity and therefore better nutritional balance. Another theory is that it could prevent predation being focused on one species only, which could result in depletion and reduced availability of prey in the locality.

In comparison, other cats may continue to eat only one type and texture of food and reject all others. This is a behaviour that might begin even before the kitten is born. At the time of weaning, kittens will prefer the taste of food eaten by their mother while she was pregnant and during lactation (see Chapter 5). If the mother has a limited diet during this time, and the kittens are not offered a variety of other food during and soon after weaning, they will be more likely to reject new food tastes and textures as adults.

Feeding patterns

Many cats prefer to eat ad-lib or little and often, which mimics the natural feeding pattern of wild and feral cats that are most likely to kill and eat several small prey animals at varying intervals throughout the day and night. Some cats will, however, eat all the food in their bowl at once, especially if it is of very high value to them; for example, if it is highly palatable or if the cat has an increased appetite owing to disease or malnutrition. Another reason for clearing the plate at one sitting can be real or perceived competition for food with other animals, especially other cats within a multi-cat household. This can be more likely to occur when cats are fed in close proximity to each other.

Predatory Behaviour

Cats have an undeserved reputation for being 'cold-blooded' killers that prey on wildlife purely for 'fun'. But this is a highly anthropomorphic assessment of feline behaviour that does not take into account the cat's instinctive survival skills. Because cats are obligatory carnivores it means that, for unowned cats especially, hunting is necessary for survival.

Well-fed and well-cared-for pet cats generally have a reduced propensity for hunting (Silva-Rodríguez and Sieving, 2011). But a strong motivation to hunt, at least occasionally, can still exist (Adamec, 1976). This is because the incentive to hunt, although increased by hunger, is not driven by hunger alone (Biben, 1979). This is a necessary survival strategy because if a cat were to wait until it is hungry to start hunting it could run a serious risk of starvation for the following reasons:

- Not all hunting attempts are successful. On average only around 50% result in a kill.
- Suitable prey is not always available.
- Hunting requires energy; a hungry cat might lack sufficient energy and might therefore be less successful.

Cats are also often accused of not killing their prey quickly and needlessly 'playing' with it. But there are equally good reasons for this behaviour:

- Rodents particularly, can inflict painful and injurious bites. To avoid injury it might be necessary for the cat to exhaust the prey sufficiently before attempting capture.
- The prey needs to be manoeuvred into the correct position so that the cat can inflict the final killing bite effectively. If the prey is large or very active this might only be possible if it is sufficiently subdued.

The hunt and kill

Cats are supremely designed to hunt, with much of their anatomy and physiology enabling them to be highly efficient predators (Box 4.1).

Prey is located initially by sound. Once the prey is located, the cat approaches fairly quickly, running while maintaining a crouching posture. It will then normally stop a short distance away and drop to the ground to observe the prey. Adopting the following position in readiness to spring:

- Body pressed flat to the ground.
- Fore legs positioned beneath the shoulders with feet flat to the ground.
- Head and neck stretched forward.
- Ears erect and forward facing.
- Eyes fully focused on the prey (Fig. 4.5).

As the cat prepares to spring, the hindquarters are raised slightly off the ground and the cat starts to tread alternately with the hind feet. Tail twitching is another common movement during hunting.

In situations where the cat might be less able to maintain good visualization of the exact prey location, e.g. in long grass, the cat may leap vertically onto the prey. But more often the cat springs forward into a short run, then either grabs the prey immediately in the mouth or initially traps it with the front feet and then grabs it in the mouth. Once a kill has been made the cat may eat it straight away or take it to a secure place away from potential competition. For pet cats this can often be the home that it shares with its human owners.

> **Box 4.1. The design of the perfect predator.**
>
> **Hearing**
>
> - Encompasses the range of very high-pitched sounds made by small mammals and birds.
> - Sensitive enough to hear the movements of small animals.
> - The pinna (outer part of the ear) can be moved independently to act as a directional amplifier to help locate the sounds of prey precisely.
>
> **Sight**
>
> - Eyes are large and forward facing, allowing 3D binocular vision, enabling the cat to pinpoint correctly the position of prey.
> - An enhanced ability to detect small, fast movements.
> - Increased ability to see in low light conditions, allowing effective hunting when nocturnal and crepuscular prey animals are most active and more likely to be out of safe hiding places.
>
> **Teeth**
>
> - Canine teeth are long and thin and used for dislocating the vertebrae of prey, enabling a quick kill.
> - Mechanoreceptors contained within the teeth allow precise location of where the killing bite should be made.
>
> **Claws**
>
> - Sharp and pointed.
> - Protractile: can be kept retracted until required.
> - Mechanoreceptors at the base of the claws enable the cat to feel and grab its prey with the correct amount of pressure.

Fig. 4.5. Hunting, preparing to pounce.

Effects on wildlife

A cat's hunting prowess, although highly valued by people that keep or encourage cats expressly for this purpose, is more likely to be considered an undesirable trait by most pet owners. Free-ranging cats, pet, stray or feral, are particularly unpopular with conservationists and wildlife enthusiasts, which is understandable because domestic cats can be efficacious predators of small mammals, birds and reptiles and are reported to have been at least partly responsible for the extinction of several endangered species, and may pose a serious threat to many more (Medina *et al.*, 2011; Thomas *et al.*, 2012). There is, however, some debate as to whether cats are sometimes given disproportionate blame for changes in wildlife populations and that other factors, such as habitat disruption and predation by wild animals, are sometimes disregarded or overlooked (Bradshaw, 2013).

The impact that cats have on wildlife populations can depend greatly on the environment. Unique and isolated fauna on small islands can be at most risk from introduced cats and the eradication of feral cats from some islands has resulted in the dramatic recovery in the populations of previously threatened species (Cooper *et al.*, 1995). However, cats can also help to effectively control the population of rats and other introduced species that can pose a threat to wildlife or their habitat. In some instances, the eradication of cats has put endangered species at greater rather than reduced risk (Karl and Best, 1982; Dickman, 2009). For example, on one Pacific island the removal of cats resulted in a dramatic increase in the rabbit population, resulting in severe damage to the environment and devastating consequences for the indigenous wildlife and nesting seabirds (Bergstrom *et al.*, 2009).

References

Adamec, R.E. (1976) The interaction of hunger and preying in the domestic cat (*Felis catus*): an adaptive hierarchy? *Behavioral Biology* 18, 263–272.

Allaby, M. and Crawford, P. (1982) *The Curious Cat*. Michael Joseph, London.

Bartoshuk, L.M., Jacobs, H.L., Nichols, T.L., Hoff, L.A. and Ryckman, J.J. (1975) Taste rejection of nonnutritive sweeteners in cats. *Journal of Comparative and Physiological Psychology* 89, 971.

Bergstrom, D.M., Lucieer, W., Kiefer, K., Wasley, J., Belbin, L., Pedersen, T.K. and Chown, S.L. (2009) Indirect effects of invasive species removal devastate World Heritage Island. *Journal of Applied Ecology* 49, 73–81.

Biben, M. (1979) Predation and predatory play behaviour of domestic cats. *Animal Behaviour* 27, 81–94.

Bowlby, J. (1988) *A Secure Base. Clinical Applications of Attachment Theory*. Routledge, Abingdon, UK.

Bradshaw, J.W.S. (2013) *Cat Sense: The Feline Enigma Revealed*. Allen Lane, Penguin Books, London.

Bradshaw, J.W.S. (2016a) Sociality in cats: a comparative review. *Journal of Veterinary Behavior* 11, 113–124.

Bradshaw, J.W.S. (2016b) What is a cat, and why can cats become stressed or distressed? In: Ellis, S. and Sparkes, A. (eds) *The ISFM Guide to Feline Stress and Health*. International Cat Care, Tisbury, Wiltshire, UK.

Bradshaw, J.W.S. and Hall, S.L. (1999) Affiliative behaviour of related and unrelated pairs of cats in catteries: a preliminary report. *Applied Animal Behaviour Science* 63, 251–255.

Bradshaw, J.W.S., Casey, R.A. and Brown, S.L. (2012) *The Behaviour of the Domestic Cat*, 2nd edn. CAB International, Wallingford, UK.

Cole, D.D. and Shafer, J.N. (1966) A study of social dominance in cats. *Behaviour* 27, 39–52.

Cooper, J., Marais, A.V.N., Bloomer, J.P. and Bester, M.N. (1995) A success story: breeding of burrowing petrels (Procellariidae) before and after the eradication of feral cats *Felis catus* at subantarctic Marion Island. *Marine Ornithology* 23, 33–37.

Corbett, L.K. (1979) Feeding ecology and social organisation of wildcats (*Felis silvestris*) and domestic cats (*Felis catus*) in Scotland. PhD thesis, University of Aberdeen, Aberdeen, UK.

Crowell-Davis, S.L., Curtis, T.M. and Knowles, J. (2004) Social organization in the cat: a modern understanding. *Journal of Feline Medicine and Surgery* 6, 19–28.

Curtis, T.M., Knowles, R.J. and Crowell-Davis, S.L. (2003) Influence of familiarity and relatedness on proximity and allogrooming in domestic cats (*Felis catus*). *American Journal of Veterinary Research* 64, 1151–1154.

Dards, J.L. (1983) The behaviour of dockyard cats: interactions of adult males. *Applied Animal Ethology* 10, 133–153.

Dickman, C. (2009) House cats as predators in the Australian environment: impacts and management. *Human–Wildlife Conflicts* 3, 41–48.

Edwards, C., Heiblum, M., Tejeda, A. and Galindo, F. (2007) Experimental evaluation of attachment behaviors in owned cats. *Journal of Veterinary Behavior* 2, 119–125.

Geering, K. (1986) Der Einfluss der Fütterung auf die Katze-Mensch-Beziehung. Thesis, University of Zürich-Irchel, Switzerland. Cited in: Turner, D.C. (2000) The human–cat relationship. In: Turner, D.C. and Bateson, P. (eds) *The Domestic Cat: The Biology of its Behaviour*, 2nd edn. Cambridge University Press, Cambridge, UK, pp. 119–147.

Horwitz, D.F., Soulard, Y. and Junien-Castagna, A. (2010) The feeding behavior of the cat. In: Pibot, P., Biourge, V. and Elliot, D.A. (eds) *Encyclopedia of Feline Clinical Nutrition*. International Veterinary Information Service, Ithaca, New York. Available at: http://www.ivis.org/docarchive/A5115.0110.pdf (accessed 18 March 2018).

Karl, B.J. and Best, H.A. (1982) Feral cats on Stewart Island; their foods, and their effects on kakapo. *New Zealand Journal of Zoology* 9, 287–293.

Kerby, G. and Macdonald, D.W. (1988) Cat society and the consequence of colony size. In: Turner, D.C. and Bateson, P.P.G. (eds) *The Domestic Cat: The Biology of its Behaviour*, 1st edn. Cambridge University Press, Cambridge, UK, pp. 67–81.

Kitts-Morgan, S.E., Caires, K.C., Bohannon, L.A., Parsons, E.I. and Hilburn, K.A. (2015) Free-ranging farm cats: home range size and predation on a livestock unit in Northwest Georgia. *PLoS One*, DOI: 10.1371/journal.pone. 0120513.

Levine, E., Perry, P., Scarlett, J. and Houpt, K.A. (2005) Intercat aggression in households following the introduction of a new cat. *Applied Animal Behaviour Science* 90, 325–336.

Liberg, O. and Sandell (1988) Spatial organisation and reproductive tactics in the domestic cat and other felids. In: Turner, D.C. and Bateson, P. (eds) *The Domestic Cat: The Biology of its Behaviour*, 1st edn. Cambridge University Press. Cambridge, UK, pp. 83–98.

Liberg, O., Sandell, M. and Pontier, D. (2000) Density, spatial organisation and reproductive tactics in the domestic cat and other felids. In: Turner, D.C. and Bateson, P. (eds) *The Domestic Cat: The Biology of its Behaviour*, 2nd edn. Cambridge University Press, Cambridge, UK, pp. 119–147.

Macdonald, D.W., Apps, P.J., Carr, G.M. and Kerby, G. (1987) Social dynamics, nursing coalitions and infanticide among farm cats, *Felis catus*. *Ethology* 28, 66.

Macdonald, D.W., Yamaguchi, N. and Kerby, G. (2000) Group-living in the domestic cat: its sociobiology and epidemiology In: Turner, D.C. and Bateson, P. (eds) *The Domestic Cat: The Biology of its Behaviour*, 2nd edn. Cambridge University Press, Cambridge, UK, pp. 96–118.

Masserman, J.H. and Siever, P.W. (1944) Dominance, neurosis and aggression: an experimental study. *Psychosomatic Medicine* 6, 7–16.

Medina, F.M., Bonnaud, E., Vidal, E., Tershy, B.R., Zavaleta, E.S., Josh Donlan, C., Keitt, B.S., Corre, M., Horwath, S.V. and Nogales, M. et al. (2011) A global review of the impacts of invasive cats on island endangered vertebrates. *Global Change Biology* 17, 3503–3510.

Murray, J.K., Roberts, M.A., Whitmarsh, A. and Gruffydd-Jones, T.J. (2009) Survey of the characteristics of cats owned by households in the UK and factors affecting their neuter status. *Veterinary Record* 164, 137–141.

Potter, A. and Mills, D.S. (2015) Domestic cats *(Felis silvestris catus)* do not show signs of secure attachment to their owners. *PloS One* 10(9), e0135109.

Prato-Previde, E., Custance, D.M., Spiezio, C. and Sabatini, F. (2003) Is the dog–human relationship an attachment bond? An observational study using the Ainsworth's strange situation. *Behaviour* 140, 225–254.

Schwartz, S. (2002) Separation anxiety syndrome in cats: 136 cases (1991–2000). *Journal of the American Veterinary Medical Association* 7, 1028–1033.

Shreve, K.R., Mehrkam, L.R. and Udell, M.A. (2017) Social interaction, food, scent or toys? A formal assessment of domestic pet and shelter cat (*Felis silvestris catus*) preferences. *Behavioural Processes* 141, 322–328.

Silva-Rodríguez, E.A. and Sieving, K.E. (2011) Influence of care of domestic carnivores on their predation on vertebrates. *Conservation Biology* 25, 808–815.

Thomas, R.L., Fellowes, M.D.E. and Baker, P.J. (2012) Spatio-temporal variation in predation by urban domestic cats (*Felis catus*) and the acceptability of possible management actions in the UK. *PLoS One* 7(1–13), e49369. DOI: 10.1371/Journal.pone.0049369.

Thomas, R.L., Baker, P.J. and Fellowes, M.D.E. (2014) Ranging characteristics of the domestic cat (*Felis catus*) in an urban environment. *Urban Ecosystems* 17(4), 911–921.

Topál, J., Miklósi, Á., Csányi, V. and Dóka, A. (1998) Attachment behavior in dogs (*Canis familiaris*): a new application of Ainsworth's (1969) Strange Situation Test. *Journal of Comparative Psychology* 112, 219–229.

Zak, P.J. (2016) Cats vs Dogs. Available at: http://www.ohmidog.com/tag/paul-zak (accessed 7 February 2018).

Zoran, D.L. (2002) The carnivore connection to nutrition in cats. *Journal of the American Veterinary Medical Association* 221(11), 1559–1567.

5 Kitten to Cat – Reproduction and the Behavioural Development of Kittens

Reproductive Behaviour of Tomcats

An unneutered, sexually mature male is known as a 'tom'. A great deal of a tomcat's behaviour is designed to increase his opportunity to mate with as many females as possible. Much of this behaviour can make an entire male a lot less desirable as a household pet.

Aggression/fighting

Entire male cats will generally try to avoid each other but the presence of an in-oestrus female can attract them to the same location. Fighting between entire males can be frequent and intense, involving a lot of prolonged loud yowling, growling and shrieking, escalating to actual fighting that often results in injury. The intensity of the aggression may reduce the more often two male cats encounter each other, especially if one wins the fights more often so that the other learns to back down and retreat sooner. Directly after a fight the winner will normally spray mark and rub on nearby inanimate objects far more than the loser.

Urine spraying

Although all cats, including neutered cats of both sexes, can and do urine mark, spraying is a behaviour that is most commonly performed by entire males, especially when they are close to or aware of a female in oestrus.

Spraying in entire cats is an important way that information regarding the sexual status and physical fitness of the cat is passed on via olfactory communication to potential mates and sexual rivals (see Chapter 3). This information is provided via pheromones produced in the urogenital region and collected by the urine during spray marking and also possibly by the chemical content of the urine, especially the chemical compound felinine. The excretion of felinine is regulated by testosterone. Entire, sexually mature male cats produce more felinine in their urine than females or castrated males (Miyazaki *et al.*, 2006). Felinine may also provide information about the fitness of the depositor because it is biosynthesized from the amino acids cysteine and methionine, both of which are found in meat, an essential part of a cat's diet and so more likely to be found in sufficient quantities in the system of a physically fit and successful hunter (Hendriks *et al.*, 1995).

Rubbing

The F2 fraction of the feline facial pheromones (see Chapter 3) has been identified as being associated with sexual marking in males. The depositing of this chemical signal and the visual display of rubbing appear to be part of the male courtship behaviour.

Vocalizing

Vocalizing, producing a distinct mowl sound, is another part of courtship behaviour by males (Dards, 1983; Shimizu, 2001). Breeders sometimes report that they can be alerted to the fact that their breeding queens are coming into oestrus by the 'singing' of their stud cats.

Mating

Mating is a somewhat rapid event in cats. The tom, approaching from either the side or from behind, grasps the back of the female's neck in his mouth and then mounts her. Alignment of the genital regions is assisted by the female lowering her chest while elevating her pelvis and deviating her tail to one side. Pelvic thrusting and ejaculation occurs very quickly, usually around 20–30 seconds following intromission.

Penile spines

The glans penis of a sexually intact tom is covered in pointed spines that develop at puberty and regress after castration and therefore appear to be linked to circulating testosterone. It has been hypothesised that these have three possible functions during mating:

- To provide additional sexual stimulation for the male and encourage pelvic thrusting (Cooper, 1972).
- To act as a 'holdfast' mechanism to prevent pre-ejaculatory withdrawal.
- To provide additional stimulation of the female to encourage ovulation (Aronson and Cooper, 1967).

Infanticide

The killing of cubs fathered by males from another pride is a recognized behaviour of male lions. But, although the behaviour has been observed in domestic cats (MacDonald *et al.*, 1987; Pontier and Natoli, 1999), it is not clear how common it is or whether these observations were examples of aberrant behaviour. Killing a female's kittens would have the effect of bringing the queen back into oestrus quicker. It is therefore possible that a tom with poor mating success might use this strategy to increase his chances of mating. Nursing queens can sometimes be more aggressive than normal to males, especially to unknown males. The possibility that they might pose a significant risk to her kittens might explain why this is.

Reproductive Behaviour of the Queen

An unneutered, sexually mature female is known as a queen. The usual age of puberty is around 6–9 months, but a female may have her first season as early as 4 months of age. Others may not start their reproductive cycle until they are almost a year old. The age of puberty can vary between breeds but also the time of year that a female cat is born may influence the timing of her first season.

Seasonal polyoestrous

Cats are seasonal breeders, which means that they only enter their reproductive cycle at a certain time of year. Female cats have multiple oestrus cycles each year that usually begin in spring as daylight starts to lengthen and end in late autumn to early winter as daylight shortens. During the winter months when daylight length is reduced they normally enter anoestrus, a period of hormonal inactivity.

Some breeders anecdotally report that their queens also go into oestrus during the winter months and studies of free-ranging cats have observed pregnancies throughout the year. The number of pregnancies identified are, however, notably fewer than are seen during the spring and summer months (Prescott, 1973; Nutter et al., 2004).

Induced ovulation

Cats are also induced ovulators. This means that mating, which triggers a release of luteinizing hormone from the pituitary, is required to stimulate the release of eggs from the ovarian follicles, something that occurs spontaneously in other mammals such as humans and dogs.

Numerous matings may be needed for sufficient luteinizing hormone to be released, and a free-ranging queen might be mated by several males over a period of 24–48 hours. Natoli and De Vito (1988) observed that the free-ranging females in their study mated with up to seven males; however, the number of males with which a female mates can depend on the local density of intact, sexually mature toms. Also, not all mating attempts may be accepted. Some females employ a mate choice, only allowing herself to be mated by the males she chooses. A possible reason for this might be to avoid inbreeding (Ishida et al., 2001).

If ovulation during oestrus is not stimulated, either because mating did not occur or was insufficient to release sufficient luteinizing hormone, the queen will then enter interoestrus when she is no longer sexually receptive for a period of around 7–9 days (range 2–19 days) before going back into oestrus or anoestrus, depending on the time of year.

If ovulation does take place she will enter a period of metoestrus, which usually encompasses pregnancy. But if ovulation occurs without resulting in pregnancy the metoestrus phase may then be considered a period of false or pseudopregnancy lasting approximately 30–45 day before going back into oestrus (Table 5.1) (Paape et al., 1975; Verhage et al., 1976).

Table 5.1. Stages and associated behavioural signs of the feline oestrus cycle.

Stage	Duration	Hormonal and physiological changes	Behavioural signs
Pro-oestrus	1–2 days	The follicles containing the eggs develop under the influence of luteinizing hormone (LH) and follicle stimulating hormone (FSH) causing the secretion of oestrogen from the ovarian follicles	May show no change in behaviour, or may start to demonstrate behaviour associated with oestrus. Males may be interested in the female but she will not yet be receptive to mating
Oestrus	2–19 days	Plasma oestrogen increases. Female becomes receptive to mating.	Calling: distinctive, persistent and loud vocalizations. Increased rubbing on inanimate objects to deposit scent from facial glands. Urine marking. Rolling and purring. Lateral deviation of the tail. Lordosis: downward arching of the back with hindquarters raised. The last three behaviours are more likely to be seen when in the presence of the male
Interoestrus	2–17 days. Average 7–9 days before returning to oestrus	A period of reproductive inactivity that occurs if ovulation has not taken place during oestrus	Returns to normal behaviour. Does not attract males
Metoestrus pregnancy	Gestation periods can vary between breeds but is on average 65–67 days		Individual variation in behaviour – some queens become more docile and affectionate. Some breeders also anecdotally report an occasional increase in agitated, anxious and aggressive behaviour

Continued

Table 5.1. Continued.

Stage	Duration	Hormonal and physiological changes	Behavioural signs
Metoestrus pseudopregnancy	Average 30–45 days before returning to oestrus	Occurs if ovulation takes place without conception. Progesterone levels increase, but not to the same level as in pregnancy and decline much earlier	May show the same behavioural changes associated with early pregnancy, although most show very little or no physical or behavioural changes
Anoestrus	Variable, can be several months	A period of no reproductive activity, usually the winter months	Returns to normal behaviour. Does not attract males

Mating

A female that is ready to mate will adopt a position known as lordosis whereby she lowers her front end, raises her pelvis and moves her tail to one side. This position signals to the male that she is ready to mate and it also helps the male to achieve intromission. Very soon after intromission she will usually emit a loud high-pitched cry followed by turning aggressively towards the male. She will then vigorously roll, stretch, and lick her external genitalia.

Pregnancy

The normal gestation period can vary between breeds and individuals but it generally lasts 64–68 days, approximately 9 weeks. This is a few days longer than the average pregnancy of the cat's wild ancestor *Felis lybica*. Physical and behavioural changes are normally seen during pregnancy (see Box 5.1).

Birth

A pregnant cat does not build a nest but she will invest a great deal of time and effort in finding the most suitable and safe area to give birth. This will usually be somewhere enclosed and sheltered that is also close to essential resources such as food and water (Lawrence, 1980). Once the nest site has been chosen she will rub around the area to deposit her scent.

Signs of imminent parturition

- Decreased appetite (she may even refuse all food immediately prior to giving birth).
- Increased vocalization.
- Panting.
- Slight reduction in body temperature.

- Increased self-grooming, especially the anogenital area and around the nipples. Newborn kittens use olfactory (scent) cues to help them locate the teats and the mother's saliva might provide an additional scent that helps to attract them to the correct area to feed (Raihani et al., 2009; Bradshaw et al., 2012).

> **Box 5.1. Changes during pregnancy**
>
> **Physical changes**
>
> Weight gain – usually around 1–3 lbs (0.5–1.5 kg) by the end of pregnancy.
> Abdominal swelling.
> Increased appetite.
> Vomiting, but this does not happen with all pregnancies and if vomiting is frequent or prolonged veterinary advice must be sought.
> Swelling and reddening of the nipples from around 3–4 weeks into gestation. This is known as 'pinking up'. Rapid swelling of the mammary glands during the final week of pregnancy.
>
> **Behavioural changes**
>
> Reduction in activity and agility, increased sleeping.
> Alterations in behaviour towards people and other animals may be seen during late pregnancy, either becoming more docile and friendly and seeking more contact and attention or showing an increase in agitated and aggressive behaviour. This might not be a complete change in behaviour but an exaggeration of the cat's normal temperament and her relationship with other individuals.

Parturition (giving birth – sometimes referred to as 'queening')

- During the first stage of labour the queen starts breathing through her mouth, becomes increasingly restless and continues to groom herself. Purring is also very common. This stage might only last a few hours or might continue for 24 hours or more.
- The kittens are delivered during stage two of labour. The time between delivery of each kitten is usually around 30–60 minutes but it can be up to 2 hours.
- Stage three is the expulsion of the placenta. This may occur immediately after each kitten is born or it might be passed along with the next kitten.
- As each kitten is born the queen cleans away the birth membranes and cuts the umbilical cord with her teeth. Both the umbilical cord and placenta are usually eaten. This not only helps to keep the nest site clean and decreases the risk of attracting predators, it also provides her with a source of nutrition. Although this is not necessary for a well-fed pet, it can be essential for a wild or feral cat that, in the first few days after giving birth, may have limited opportunity to leave the nest to hunt. Occasionally an inexperienced or incompetent mother might not be able to differentiate between a newborn kitten and the placenta and may kill and eat the kitten as well (Baerends-van Roon and Baerends, 1979, cited in Deag et al., 2000).

Pre-weaning period

Immediately after the kittens are born the mother will usually lie on her side, encircling the kittens, providing them with warmth and allowing them access to her nipples, occasionally adjusting her position as necessary to suit them. She is likely to remain in this position, frequently grooming and nuzzling the kittens for the first 24 hours or so. Later she may assume other nursing positions such as a half-sitting posture. At this stage it is estimated that she will spend around 70% of her time with her kittens (Bradshaw *et al.*, 2012).

It is important that mother and kittens are disturbed as little as possible during the first week or so after birth. Severe or frequent disturbance may result in the mother abandoning or even attacking the kittens. Other reasons for abandonment or killing of newborn kittens can include:

- Immature mother. A queen that becomes pregnant while she is still very young may not have the physical or behavioural maturity to cope with motherhood.
- Unwell mother. If the mother cat is unwell or undernourished this might affect her behaviour and she might also be unable to feed the kittens. Mastitis can also make it very painful for the mother cat to nurse her kittens.
- Sickness or deformity of the kittens. A queen might abandon kittens that are unhealthy or have physical deformities.

It is normal, however, for the mother to move the kittens to a new nest site at least once during the pre-weaning period (Fig. 5.1). There have been a few hypotheses put forward to explain this behaviour, including the need to find a clean site if the nest has become dirty or infected by parasites (Corbett, 1979; Deag *et al.*, 2000); however,

Fig. 5.1. It is normal for the mother to move the kittens to a new nest site at least once during the pre-weaning period.

Feldman (1993) found very little evidence to support this. Other reasons that have been postulated are disturbance by other cats, especially unknown males, or the need to move to a new site where prey may be more plentiful (Fitzgerald and Karl, 1986; Deag *et al.*, 2000). Moving the kittens will also help to keep them hidden from potential predators (Bradshaw *et al.*, 2012).

Maternal aggression

During and after parturition a queen may become aggressive towards conspecifics and other animals, including humans that she may have previously tolerated or appeared to be friendly with. This behaviour is more likely to occur if she has reason to believe that her kittens might be under threat. Therefore, care should be taken to avoid severe disturbance and too frequent or inappropriate handling of the kittens.

It is also advisable that queens that have demonstrated maternal aggression are not bred from again because this behaviour can have a detrimental influence on the future behaviour and welfare of her kittens. Occasionally this aggressive behaviour continues after the kittens have been weaned, especially if the kittens remain with the mother, but it usually declines after weaning.

Weaning

When the kittens are around 3 weeks of age the mother starts the weaning process by bringing to the nest site initially dead and then later live, but injured or exhausted prey for them to play with, kill and eventually eat. Once they have mastered capturing and killing weakened prey, she will then provide them with live, active and uninjured prey (Fig. 5.2).

Fig. 5.2. The mother starts the weaning process by bringing injured or exhausted prey to the nest site for the kittens so that they can learn and practise their hunting skills.

With pet cats this process is often disrupted by human owners and cannot occur at all if the mother is confined indoors. In this case, she might bring toys and other items to her kittens to play with and practise their hunting skills.

Cats are more likely to continue to hunt and kill the same species as their mother (Caro, 1980). This is most likely due to their initial experience with prey but it has also been found that what a queen eats during her pregnancy can influence her kittens' food preferences, even more so if she continues to eat the same food when nursing (Becques *et al.*, 2009; Hepper *et al.*, 2012). This could be a survival adaptation in that the kittens will be more likely to eat things that are known and therefore safe. It can also be a great help when weaning pet kittens if the first food they are offered is the same flavour as the food the mother has been eating whilst pregnant.

Frustration during weaning

Part of the natural weaning process involves the mother spending a gradually increasing amount of time away from the kittens as they mature. When she does allow them to nurse she may often get up and move away before they have finished feeding. This is probably because nursing becomes increasingly uncomfortable for her because of the kittens developing teeth, plus their increased strength and vigour when suckling. The result is that the kittens regularly experience frustration, which encourages them to search for other food sources and so aids the weaning process. Early attempts at capturing and killing prey brought to the nest by the queen can also involve an element of delayed reward and associated frustration for the kittens, although the mother will normally despatch the prey eventually if the kittens take too long to kill it themselves.

These early experiences are an important part of behavioural development because they provide the kittens with the opportunity to learn how to cope with the emotional state of frustration. In contrast, when kittens are hand-reared, conscientious human carers may continue to feed the kittens 'on demand' or at predictable intervals, and supply sufficient milk so that they are rarely left unsatisfied. Also, when solid food is introduced in a bowl it requires no effort by the kitten to obtain it. Therefore, hand-raised kittens may have very little experience of frustration at this early stage in their life and therefore might not get the opportunity to learn how to cope with the emotion, which can lead to problem behaviour such as frustration-related aggression in later life (Bowen and Heath, 2005).

Physical and Behavioural Development of Kittens

Influences on behaviour in utero (before birth)

Inherent factors

Personality traits of confidence or timidity and overall friendliness can be at least partially inherent. McCune (1995) found that kittens fathered by confident males were far quicker to approach, investigate and interact with novel objects and unfamiliar people than those fathered by less confident males. Turner *et al.* (1986) found a similar effect when laboratory cats were tested regarding their friendliness towards familiar people. In both studies, as is usual for cats, the father had no contact at all with the kittens so this influence can only be genetic.

Prenatal stress

Described briefly, stress is an emotional and physiological reaction to a physiological, psychological or emotional challenge. The stress response is an important and necessary function because it increases the ability to escape from or defend against danger. But the chemical changes within the body that stress causes, particularly the production of 'stress hormones', can have detrimental effects on both health and welfare if produced excessively or long-term (see Chapter 6).

Prenatal stress is stress experienced by the mother during pregnancy, which has been shown to have physiological and behavioural effects on the developing foetus (Weinstock, 2008).

EPIGENETICS. Literally meaning 'above genetics', epigenetics relates to how the expression of DNA, but not the DNA sequence, can be altered by external influences (Jensen, 2013). Stress hormones produced by the mother during pregnancy can cross the placenta and cause epigenetic changes to her developing offspring, altering how they deal with stress in later life. This can be advantageous for the kittens if the level of stress they encounter throughout life matches that experienced by their mother during pregnancy but, if not, then their stress response will be excessive and maladaptive to their environment. This can result in individuals that show signs of increased emotionality, including anxiety, fear and reactivity, which in turn can be linked to increased aggression and antisocial behaviour (Simonson, 1979, cited in Bateson, 2000).

THE NUTRITIONAL STATUS OF THE MOTHER. Undernourishment of the mother is one cause of prenatal stress but it can also have other critical influences. Smith and Jansen (1977) (cited in Bateson, 2000) found severely undernourished mothers gave birth to kittens with growth defects in the brain, resulting in abnormalities in both motor and behavioural development. Further studies have also found that kittens born to mothers subjected to insufficient nutrition during pregnancy show increased emotionality, reduced learning ability, poor motor skills and are more likely to be aggressive towards other cats (Simonson, 1979, cited in Bateson, 2000; Gallo *et al.*, 1980).

Postnatal physical development

The motor skills and senses are far from fully developed at birth (Fig. 5.3) but a kitten's tactile, thermal and olfactory senses are sufficiently well developed for these to

Fig. 5.3. Cats are an altricial species, meaning that their senses and motor skills are undeveloped at birth and they are fully reliant on the mother cat for food, warmth and protection.

dominate its sensory world for the first 2 weeks of life. Newborn kittens have full pain perception and can sense temperature and will actively move away from cold and towards warmth (Raihani et al., 2009; Bateson, 2014). This is an essential ability for the newborn because they are not able to regulate their own body temperature for another 3–4 weeks (Olmstead et al., 1979; Bradshaw, 1992; Lawler, 2008).

Sense of smell

Olfaction (sense of smell) is present and reasonably well developed at birth but it is not fully mature until around 3 weeks of age. It is, however, sufficiently well developed to allow a kitten to locate a nipple of a lactating female by touch and scent alone. The 'gape' or 'flehmen response', indicating usage of the vomeronasal or 'Jacobson's organ' (see Chapter 2), is not seen until 5 weeks of age and not fully developed until the kitten is at least 7 weeks old.

Hearing

Although audition (hearing) is reasonably well developed at birth the newborn kitten has very limited hearing because the ear canals are blocked by ridges of skin. As the kitten grows and the ear canals widen, these ridges decrease and hearing gradually improves. The first responses to sound are usually seen at around 5 days after birth and by about 16 days the canals have widened sufficiently, enabling most kittens to locate the source of a sound. Hearing continues to develop and by 3–4 weeks of age most kittens can discriminate between different feline vocalizations, being more likely to approach familiar and friendly calls made by the mother and littermates, while showing fear and attempting to move away from growls and other threatening feline calls, especially those made by unknown cats (Olmstead and Villablanca, 1980; Bradshaw, 1992).

Sight

Vision is the sense that takes the longest time to develop. The average age at which the eyes open is 7–10 days after birth but this can vary from 2 to 16 days (Villablanca and

Olmstead, 1979, cited in Martin and Bateson, 1988). The variation in eye-opening age seems to be due to several possible factors. Females generally tend to open their eyes earlier than males and kittens that receive an increased amount of handling are more likely to open their eyes sooner than those that are less well handled. There also seems to be a paternal influence (Bradshaw *et al.*, 2012). Other factors such as the level of ambient light and the age of the mother also seem to be significant, with kittens raised in reduced light conditions and those born to young mothers opening their eyes sooner (Braastad and Heggelund, 1984).

Visual orienting and following normally develop at around 15–25 days, and obstacle avoidance at around 25–35 days. Vision is usually sufficiently functional by around 5 weeks of age but gradual improvements in visual acuity (clarity of vision) continue until the kitten is around 10–16 weeks (Ikeda, 1979; Sireteanu, 1985; Martin and Bateson, 1988).

A cat's resting eye state, whether it is long- or short-sighted, seems to depend on early visual experience. Indoor-raised cats are more likely to be short-sighted and feral cats or those raised with ample outdoor access more likely to be long-sighted (Hughes, 1972).

Motor development

Newborn kittens are not completely immobile, but during the first 2 weeks of life their movements are uncoordinated, and they can only move forward using a slow and ineffective paddling gait. The ability to return to an upright position, the body-righting reflex, is present from birth although not fully developed for another month.

Forelimb coordination develops during the first 2 weeks after birth and hind-limb coordination sometime within the following 2 weeks. Most kittens can stand by the time they are 10 days old and are able to walk by 2–3 weeks of age but are still somewhat uncoordinated. It is not for at least another week or so that they are able to move any significant distance. At around 5 weeks they can run for very short distances and move quite freely by 6–7 weeks of age, although the full complex balancing skills seen in the adult cat do not yet become fully developed for another few weeks (Martin and Bateson, 1988; Bradshaw *et al.*, 2012).

Voluntary elimination develops at around 3–4 weeks, until which time the mother needs to stimulate urination and defecation by licking the anogenital region. The anogenital reflex, emptying the bladder and bowels in response to external stimulation, disappears at around 4–5 weeks of age (Thor *et al.*, 1989).

Behavioural development

Kittens demonstrate an ability to learn as soon as they are born. A preference for suckling from a particular nipple develops very soon after birth and kittens can find the same nipple, probably by following scent cues, every time that they feed (Raihani *et al.*, 2009). However, learning that can have the most influence on their later behaviour begins at around 2 weeks of age when their developing senses allow them to become more aware of their surroundings. This time signifies the beginning of the sensitive period.

The sensitive period

This is a time in an animal's life when external experience can have the greatest influence on later behaviour. In cats, this period starts at around 2 weeks of age when the kitten not only becomes sufficiently aware of its surroundings but it is also a time when neural development and increased plasticity enhances the ability to learn (Knudsen, 2004). This increased sensitivity to learning starts to decline at around 7 weeks of age, which may also be related to the emergence of adult-type escape and fear responses at around 6–8 weeks of age (Kolb and Nonneman, 1975).

The sensitive period is not the only time when learning takes place, but the experiences that a kitten is exposed to during this time can have the greatest influence in shaping its behaviour for the rest of its life. The two learning processes that are of most importance during this time are socialization and habituation:

- **Socialization** is the process whereby a young animal develops social attachments and learns to recognize and to develop appropriate social behaviour with members of its own species and with members of other species that it lives with.
- **Habituation** is the process whereby an animal learns to ignore non-threatening events and stimuli.

Cats that have not experienced sufficient or appropriate socialization with people or habituation to household noises, etc. during the sensitive period are likely to be unsuitable as domestic pets.

Socialization to cats, other species and people

A kitten's early experience can affect how it interacts with other cats as an adult. Cats that have had limited socialization with conspecifics are more likely to be fearful and potentially aggressive towards other cats and demonstrate signs of distress if confined with other cats later in life. Cats with a history of positive experience with other cats during the sensitive period generally appear less stressed and more tolerant around other cats as adults (Kessler and Turner, 1999).

Because the sensitive period starts to decline before pet kittens go to their new homes, the only other cats that many kittens are exposed to are its mother and littermates. Studies have shown that close-bonded relationships are far more likely to exist between littermates that remain together than between non-related cats that live in the same household (Bradshaw and Hall, 1999). It is unclear, however, if this is simply because of the development of attachments during the sensitive period, regardless of relatedness, or if being related increases the likelihood of continued attachment.

Experimenters have successfully raised and socialized kittens with other species, even with animals that would normally be regarded as either prey or predator. In studies where kittens were raised with rats, they grew up into cats that did not regard similar rats as prey, although they would still attack rats of other strains (Kuo, 1930). In another experiment, where kittens were raised with puppies, at 12 weeks of age the kittens showed no fear and would happily play with other puppies, whereas kittens of the same age with no experience of puppies showed fear, avoidance and defensive behaviour (Fox, 1969).

Karsh and Turner (1988) found that if a kitten has no, or very little, handling prior to 7 weeks of age it is far more likely to remain fearful of people, despite later attempts at socialization. Enhanced handling up until 9 weeks of age appears to produce even better results in the form of friendlier and less fearful cats as adults (Casey and Bradshaw, 2008).

The amount of handling is also important, with those handled for a total of at least 30–40 minutes per day, showing greater confidence and friendliness than those handled for only 15 minutes overall daily (see Chapter 8). However, unpublished data recorded by Bradshaw and Cook, cited in McCune et al. (1995), found that kittens handled for 5 hours daily were not notably different in their friendliness to people than kittens handled for just under 1 hour a day.

The number of people the kitten is handled by can also make a difference. Karsh and Turner (1988) found that those handled by at least four different people became generally well socialized to people, whereas kittens handled by only one person would become attached and show more social behaviour towards that one person but act negatively towards other people they had not been handled by, even people with whom they were otherwise familiar. To increase the likelihood that kittens will learn to feel confident and relaxed around people in general, it is also important that they are appropriately handled by a variety of people of both sexes, varying ages, and even wearing different clothes and perfumes (Karsh and Turner, 1988).

The type of handling is another thing that can influence the success of socialization. For example, speaking whilst stroking the kitten can aid the development of a relationship (Moelk, 1979), although individual differences should always be considered. For some kittens engaging in object play might be a better way to socialize and make positive associations with people (Turner, 1995).

Habituation

When an animal encounters a new event or experience, it is normal for it to be wary and to act or be prepared to act in a defensive manner (e.g. fight or flight). But to react defensively to all or most stimuli would be physically, mentally and emotionally exhausting. Habituation is a learning process whereby the animal comes to regard common and unthreatening stimuli as irrelevant and therefore has no need to attend to or react to them.

Habituation is achieved by repeated presentation of a harmless stimulus (sight, sound, smell or experience) at a level that the animal can easily adapt to and learn to ignore. The animal will also learn to habituate to the stimulus even at a heightened level, so long as the initial increase in the stimulus is gradual and not sudden or unexpected. But, if the stimulus is initially presented at too high a level, the animal will be less likely to habituate to it and may even become sensitized, i.e. become more fearful and reactive.

For example, most cats will ignore sounds from the radio or television. This is because initial experiences of television or radio sounds are likely to be at a low to moderate level and in many houses the radio and/or television are on for a few hours every day. Even if later the volume is occasionally increased, the cat is likely to consider the sound as irrelevant and continue to ignore it.

In comparison, many pet cats are frightened of the vacuum cleaner and this is a fear that often increases rather than decreases. In other words, the cat becomes sensitized to it. This is because a cat's initial experience of the noise of the vacuum cleaner is something that is sudden, unexpected and loud.

Habituation can take place at any time throughout life, but it occurs more readily during the sensitive period.

The influence of the mother and littermates during the sensitive period

The presence of the mother and siblings can aid socialization and habituation by providing the kitten with a 'safe base' from which to explore and learn about new encounters and experiences. Kittens separated from their mother and littermates will be more cautious and less willing to explore and learn (McCune *et al.*, 1995).

Kittens learn by observation of others but most especially from watching their mother. Chesler (1969) found that kittens who watched their mother perform the task of pressing a lever to get food developed the skill, not only faster than others left to work it out for themselves, but also faster than kittens who had observed an unfamiliar female cat perform the same task.

Social learning by observation of the mother's reaction can also teach the kitten who or what is potentially threatening (Gray, 1987). If the mother is calm and well socialized with people, her presence can facilitate a good human/kitten relationship (Karsh and Turner, 1988). But if a mother cat demonstrates fearful and defensive reactions towards people while the kittens are most sensitive to learning, they are more likely to become fearful of people themselves.

Kitten play

Play is also an important part of physical and behavioural development. Kittens engage in three types of play: social play, locomotor play and object play, which are defined mainly by the play target but also by the behaviour patterns involved.

Social play

Play with other kittens begins at around 3 weeks of age and starts to decline at around 12–16 weeks (Table 5.2; Fig. 5.4) (West, 1974; Caro, 1981). Social play is generally between littermates but can also be directed towards the mother cat, especially by single kittens without siblings. This can, however, sometimes result in increased maternal aggression towards the kitten and reduced maternal care, possibly owing to the mother being reluctant to engage in the more vigorous elements of social play with her kitten (Mendl, 1988).

Social play may be loosely described as 'play fighting' because it does seem to contain attenuated and moderated forms of agonistic social behaviour, and as kittens get older social play might occasionally escalate into actual fighting (Voith, 1980, cited in Bateson, 2000). Visual signals such as a half-open mouth that often accompanies social play in kittens may be indicators to the other cat or kitten that the intention of the behaviour is play and not aggression (Bradshaw *et al.*, 2012).

Table 5.2. Social play signals in kittens.

Play signal	Age first seen	Description
Belly up (Fig. 5.4a)	21–23 Days	Kitten lies on its back using all four feet to paw at or 'fight with' another kitten standing over it
Stand up (Fig. 5.4a)	23–26 Days	Standing over another kitten, pawing at or directing inhibited bites towards the head or neck of the other kitten
Side-step (Fig. 5.4b)	32–34 Days	Walks sideways with back arched and tail curved upwards, usually with focus of attention towards another kitten
Pounce (Fig. 5.4c)	33–35 Days	Initially crouches and then pounces upon or towards another kitten. May be used to initiate play
Chase (Fig. 5.4c)	38–41 Days	Running after or away from another kitten
Horizontal leap	41–46 Days	Kitten side on to another kitten; arches back, curves tail upwards and leaps up
Face-off (Fig. 5.4d)	42–48 Days	Sits near another kitten and bats or swipes at the other kitten

From West, 1974.

Fig. 5.4. Social play. (a) Belly-up and stand-up; (b) side-step; (c) pounce and chase; and (d) face-off.

Guyot *et al.* 1980 found that kittens raised without littermates may be more aggressive and be slower to learn social communication skills, indicating that social play with littermates is an important part of behavioural and social development.

Object play

Play with small inanimate objects, known as 'object play', starts later than social play as kittens develop sufficient eye–paw coordination to allow them to manipulate small items (Box 5.2) (Bateson, 2000). This seems to be partly exploratory behaviour and partly practice of predatory skills. But object play does not appear necessary for a kitten to develop the basic skills required for hunting (Martin and Bateson, 1988). Caro (1980) found that kittens with no or very limited opportunity to play with small inanimate objects demonstrated no difference in their predatory behaviour at 6 months of age than kittens that engaged in all forms of play.

> **Box 5.2. Object play in kittens**
>
> **Poke/bat**: The kitten uses a front paw to 'tap' the object from above (poke) or from the side (bat).
> **Scoop**: The kitten picks up an object by curving the front paw under it and grasping with the claws.
> **Toss**: The kitten 'throws' the object held in the mouth or grasped in the paw with a shake of the head or foot.
> **Chase**: The kitten chases a moving object, or one that has just been batted or tossed.
> **Grasp**: The kitten holds object in the mouth or between the front paws.
> **Bite/mouth**: The kitten bites or 'mouths' an object.
>
> (Adapted from Bradshaw *et al.*, 2012.)

Locomotor play

Locomotor play is playful movement or exploration that does not involve other individuals or manipulation of inanimate objects.

In a study by Martin and Bateson (1985) of seven separate litters provided with a wooden climbing frame, it was found that until the kittens were around 7 weeks of age, climbing and exploration occurred for only short periods of time and was limited to the lower rungs of the climbing frame. Even after more than 8 weeks of age some had still not ventured much higher. For those that did climb, occasionally losing balance and even falling off did not seem to be a deterrent because they would usually climb straight back up again, indicating the high reward value of the behaviour.

The mother's behaviour seemed to influence the climbing behaviour of the kittens, the most adventurous being the ones with mothers that also spent more time on the frame. Guyot *et al.* 1980 also found that kittens raised without a natural mother were more likely to be reluctant to climb when presented with ramps and shelves to explore.

Play in adult cats

Play is not just confined to kittenhood. Although adults do not play as much as kittens, playing with toys is an important part of essential environmental enrichment for adult cats.

Social play also often continues into adulthood and is a normal behaviour between cats of the same social group. But in adulthood there can be an increased tendency for some play interactions to develop into actual antagonistic encounters (Bradshaw, 1992). This may occur if arousal levels are too high or if there is already an element of competition between the cats. Another common problem can be misinterpretation by owners of inter-cat behaviour so that play is incorrectly considered to be aggression or aggression is incorrectly regarded as play.

References

Aronson, L.R. and Cooper, M.L. (1967) Penile spines of the domestic cat: their endocrine-behavior relations. *The Anatomical Record* 157(1), 71–78.

Baerends-van Roon, J.M. and Baerends, G.P. (1979) *The Morphogenesis of the Behaviour of the Domestic Cat.* Cited in: Deag, J.M., Manning, A. and Lawrence, C.E. (2000) Factors influencing the mother-kitten relationship. In: Turner, D.C. and Bateson, P. (eds) *The Domestic Cat: The Biology of its Behaviour*, 2nd edn. Cambridge University Press, Cambridge, UK, pp. 23–45.

Bateson, P. (2000) Behavioural development in the cat. In: Turner, D.C. and Bateson, P. (eds) *The Domestic Cat: The Biology of its Behaviour*, 2nd edn. Cambridge University Press, Cambridge, UK, pp. 9–22.

Bateson, P. (2014) Behavioural development in the cat. In: Turner, D.C. and Bateson, P. (eds) *The Domestic Cat: The Biology of its Behaviour*, 3rd edn. Cambridge University Press, Cambridge, UK, pp. 11–26.

Becques, A., Larose, C., Gouat, P. and Serra, J. (2009) Effects of pre- and post-natal olfactogustatory experience at birth and dietary selection at weaning in kittens. *Chemical Senses* 35, 41–45. DOI: 10.1093/chemse/bjp080

Bowen, J. and Heath, S. (2005) *Behaviour Problems in Small Animals: Practical Advice for the Veterinary Team.* Saunders, Elsevier, London.

Braastad, B.O. and Heggelund, P. (1984) Eye-opening in kittens: effects of light and some biological factors. *Developmental Psychobiology* 17, 675–681.

Bradshaw, J.W.S. (1992) *The Behaviour of the Domestic Cat.* CAB International, Wallingford, UK.

Bradshaw, J.W.S. and Hall, S.L. (1999) Affiliative behaviour of related and unrelated pairs of cats in catteries: a preliminary report. *Applied Animal Behaviour Science* 63(3), 251–255.

Bradshaw, J.W.S., Casey, R.A. and Brown, S.L. (2012) *The Behaviour of the Domestic Cat*, 2nd edn. CAB International, Wallingford, UK.

Caro, T.M. (1980) Effects of the mother, object play, and adult experience on predation in cats. *Behavioral and Neural Biology* 29, 29–51.

Caro, T.M. (1981) Sex differences in the termination of social play in cats. *Animal Behaviour* 29, 271–279.

Casey, R.C. and Bradshaw, J.W.S. (2008) The effects of additional socialisation for kittens in a rescue centre on their behaviour and suitability as a pet. *Applied Animal Behaviour Science* 114, 196–205.

Chesler, P. (1969) Maternal influence in learning by observation in kittens. *Science* 166, 901–903.

Cooper, K.K. (1972) Cutaneous mechanoreceptors of the glans penis of the cat. *Physiology & Behavior* 8, 793–796.

Corbett, L.K. (1979) Feeding ecology and social organisation of wildcats (*Felis silvestris*) and domestic cats (*Felis catus*) in Scotland. PhD thesis, University of Aberdeen, UK.

Dards, J.L. (1983) The behaviour of dockyard cats: interactions of adult males. *Applied Animal Ethology* 10, 133–153.

Deag, J.M., Manning, A. and Lawrence, C.E. (2000) Factors influencing the mother-kitten relationship. In: Turner, D.C. and Bateson, P. (eds) *The Domestic Cat: The Biology of its Behaviour*, 2nd edn. Cambridge University Press, Cambridge, UK, pp. 23–45.

Feldman, H.N. (1993) Maternal care and differences in the use of nests in the domestic cat. *Animal Behaviour* 45, 13–23.

Fitzgerald, B.M. and Karl, B.J. (1986) Home range of feral house cats (*Felis catus L.*) in forest of the Orongorongo Valley, Wellington, New Zealand. *New Zealand Journal of Ecology* 9, 71–82.

Fox, M.W. (1969) Behavioral effects of rearing dogs with cats during the 'critical period of socialization'. *Behaviour* 35, 273–280.

Gallo, P.V., Wefboff, J. and Knox, K. (1980) Protein restriction during gestation and lactation: development of attachment behaviour in cats. *Behavioral and Neural Biology* 29, 216–223.

Gray, J.A. (1987) *The Psychology of Fear and Stress,* 2nd edn. Cambridge University Press, Cambridge, UK.

Guyot, G.W., Bennett, T.L. and Cross, H.A. (1980) The effects of social isolation on the behaviour of juvenile domestic cats. *Developmental Psychobiology* 13, 317–329.

Hendriks, W.H., Moughan, P.J., Tarttelin, M.F. and Woolhouse, A.D. (1995) Felinine: a urinary amino acid of Felidae. *Comparative Biochemistry and Physiology Part B: Biochemistry and Molecular Biology* 112, 581–588.

Hepper, P.G., Wells, D.L., Millsopp, S., Kraehenbuehl, K., Lyn, S.A. and Mauroux, O. (2012) Prenatal and early sucking influences on dietary preferences in newborn, weaning and young adult cats. *Chemical Senses* 37(8), 755–766.

Hughes, A. (1972) Vergence in the cat. *Vision Research* 12, 1961–1994.

Ikeda, H. (1979) Physiological basis of visual acuity and its development in kittens. *Child Care, Health and Development* 5, 375–383.

Ishida, Y., Yahara, T., Kasuya, E. and Yamane, A. (2001) Female control of paternity during copulation: inbreeding avoidance in feral cats. *Behaviour* 138, 235–250.

Jensen, P. (2013) Transgenerational epigenetic effects on animal behaviour. *Progress in Biophysics and Molecular Biology* 113, 447–454.

Karsh, E.B. and Turner, D.C. (1988) The human–cat relationship. In: Turner, D.C. and Bateson, P. (eds) *The Domestic Cat: The Biology of its Behaviour*, 1st edn. Cambridge University Press, Cambridge, UK.

Kessler, M.R. and Turner, D.C. (1999) Socialization and stress in cats (*Felis silvestris catus*) housed singly and in groups in animal shelters. *Animal Welfare* 8, 15–26.

Knudsen, E.I. (2004) Sensitive periods in the development of the brain and behaviour. *Journal of Cognitive Neuroscience* 16, 1412–1425.

Kolb, B. and Nonneman, A.J. (1975) The development of social responsiveness in kittens. *Animal Behaviour* 23, 368–374.

Kuo, Z.Y. (1930) The genesis of the cat's response to the rat. *Journal of Comparative Psychology* 11, 1–35.

Lawler, D.F. (2008) Neonatal and paediatric care of the puppy and kitten. *Theriogenology* 70, 384–392.

Lawrence, C.E. (1980) Individual differences in the mother–kitten relationship in the domestic cat (*Felis catus*). Doctoral dissertation, University of Edinburgh, Edinburgh, UK.

Macdonald, D.W., Apps, P.J., Carr, G.M. and Kerby, G. (1987) Social dynamics, nursing coalitions and infanticide among farm cats, *Felis catus*. *Ethology* 28, 66.

Martin, P. and Bateson, P. (1985) The ontogeny of locomotor play behaviour in the domestic cat. *Animal Behaviour* 33, 502–510.

Martin, P. and Bateson, P. (1988) Behavioural development in the cat. In: Turner, D.C. and Bateson, P. (eds) *The Domestic Cat: The Biology of its Behaviour*. Cambridge University Press, Cambridge, UK, pp. 9–22.

McCune, S. (1995) The impact of paternity and early socialization on the development of cats' behaviour to people and novel objects. *Applied Animal Science* 45, 109–124.

McCune, S., McPherson, J.A. and Bradshaw, J.W.S. (1995) Avoiding problems: the importance of socialisation. In: Robinson, I. (ed.) *The Waltham Book of Human-Animal Interaction: Benefits and Responsibilities of Pet Ownership*, Waltham Centre for Pet Nutrition, Elsevier, London, pp. 71–86.

Mendl, M. (1988) The effects of litter-size variation on the development of play behaviour in the domestic cat: litters of one and two. *Animal Behaviour* 36, 20–34.

Miyazaki, M., Yamashita, T., Suzuki, Y., Saito, Y., Soeta, S., Taira, H. and Suzuki, A. (2006) A major urinary protein of the domestic cat regulates the production of felinine, a putative pheromone precursor. *Chemical Biology* 13, 1071–1079.

Moelk, M. (1979) The development of friendly approach behaviour in the cat: a study of kitten-mother relations and the cognitive development of the kitten from birth to 8 weeks. *Advances in the Study of Behaviour* 10, 163–224.

Natoli, E. and De Vito, E. (1988) The mating systems of feral cats living in a group. In: Turner, D.C. and Bateson, P. (eds) *The Domestic Cat: The Biology of its Behaviour*. Cambridge University Press, Cambridge, UK.

Nutter, F.B., Levine, J. and Stoskopf, M.K. (2004) Reproductive capacity of free-roaming domestic cats and kitten survival rate. *Journal of the American Veterinary Medical Association* 225, 1399–1402.

Olmstead, C.E. and Villablanca, J.R. (1980) Development of behavioral audition in the kitten. *Physiology and Behaviour* 24, 705–712.

Olmstead, C.E., Villablanca, J.R., Torbiner, M. and Rhodes, D. (1979) Development of thermoregulation in the kitten. *Physiology & Behaviour* 23, 489–495.

Paape, S.R., Shille, V.M., Seto, H. and Stabelfeldt, G.H. (1975) Luteal activity in the pseudopregnant cat. *Biology of Reproduction* 13, 470–474.

Pontier, D. and Natoli, E. (1999) Infanticide in rural male cats (*Felis catus* L.) as a reproductive mating tactic. *Aggressive Behavior* 25, 445–449.

Prescott, C.W. (1973) Reproduction patterns in the domestic cat. *Australian Veterinary Journal* 49, 126–129.

Raihani, G., Gonzáles, D., Arteaga, L. and Hudson, R. (2009) Olfactory guidance of nipple attachment in kittens of the domestic cat: inborn and learned responses. *Developmental Psychobiology* 51(8), 662–671.

Shimizu, M. (2001) Vocalizations of feral cats: sexual differences in the breeding season. *Mammal Study* 26, 85–92.

Simonson, M. (1979) Effects of maternal malnourishment, development, and behaviour in successive generations in the rat and cat. *Malnutrition, Environment and Behavior*. Cornell University Press, Ithaca, New York, pp. 133–160. Cited in: Bateson, P. (2000) Behavioural development in the cat. In: Turner, D.C. and Bateson, P. (eds) *The Domestic Cat: The Biology of its Behaviour*, 2nd edn. Cambridge University Press, Cambridge, UK, pp. 9–22.

Sireteanu, R. (1985) The development of visual acuity in very young kittens. A study with forced-choice preferential looking. *Vision Resolution* 25, 781–788.

Smith, B.A. and Jansen, G.R. (1977) Maternal undernutrition in the feline: brain composition of offspring. *Nutrition Reports International*. Cited in: Bateson, P. (2000) Behavioural development in the cat. In: Turner, D.C. and Bateson, P. (eds) *The Domestic Cat: The Biology of its Behaviour*, 2nd edn. Cambridge University Press, Cambridge, UK, pp. 9–22.

Thor, K.B., Blais, D.P. and de Groat, W.C. (1989) Behavioral analysis of the postnatal development of micturition in kittens. *Developmental Brain Research* 46, 137–144.

Turner, D.C. (1995) The human–cat relationship. In: Robinson, I. (ed.) *The Waltham Book of Human-Animal Interaction: Benefits and Responsibilities of Pet Ownership*. Waltham Centre for Pet Nutrition, Elsevier, London, pp. 87–97.

Turner, D.C., Feaver, J., Mendl, M. and Bateson, P. (1986) Variations in domestic cat behaviour towards humans: a paternal effect. *Animal Behaviour* 34, 1890–1892.

Verhage, H.G., Beamer, N.B. and Beamer, R.M. (1976) Plasma levels of estradiol and progesterone in the cat during polyestrus, pregnancy and pseudopregnancy. *Biology of Reproduction* 14, 579–585.

Villablanca, J.R. and Olmstead, C.E. (1979) Neurological development in kittens. *Developmental Psychobiology* 12, 101–127. Cited in: Martin, P. and Bateson, P. (1988) Behavioural development in the cat. In: Turner, D.C. and Bateson, P. (eds) *The Domestic Cat: The Biology of its Behaviour*, Cambridge University Press, Cambridge, UK, pp. 9–22.

Voith, V.L. (1980) Social play in the domestic cat. *American Zoologist* 14, 427–436. Cited in: Bateson, P. (2000) Behavioural development in the cat. In: Turner, D.C. and Bateson, P. (eds) *The Domestic Cat: The Biology of its Behaviour*, 2nd edn. Cambridge University Press, Cambridge, UK, pp. 9–22.

Weinstock, M. (2008) The long-term behavioural consequences of prenatal stress. *Neuroscience and Biobehavioral Reviews* 32, 1073–1086.

West, M.J. (1974) Social play in the domestic cat. *American Zoologist* 14, 427–436.

6 Health and Behaviour

There is a very close link between physiological and psychological welfare.

- A change in behaviour may indicate that an animal is unwell.
- Pain, discomfort and disease can negatively affect emotional well-being.
- Emotional or psychological distress can cause, trigger or exacerbate physical disease and heighten pain.

Pain

Pain is both a sensory and an emotional experience. The main function of pain is to alert the individual to actual or potential tissue damage. It can, however, also have harmful effects on welfare by negatively influencing mood and increasing stress. This is not only due to the unpleasant experience, but pain can also restrict and disrupt a cat's behavioural coping strategies by:

- Reducing the ability to escape from real or perceived threats.
- Reducing the ability to access safe and secure resting areas.
- Reducing the ability to defend and maintain territory.

Recognizing pain in cats is not easy; they can appear very stoic by nature and rarely exhibit clearly evident signs of being in pain. This is a survival tactic because revealing the presence of pain or disability could make the individual an easy target for predators or rivals.

Degenerative joint disease, which includes osteoarthritis, is one of the most common causes of pain in older cats and this can be particularly difficult to spot as it often affects cats bilaterally, making it less likely that the cat will develop an easily seen 'limp', although the cat's gait may be altered in other ways (Hardie *et al.*, 2002; Lascelles *et al.*, 2010).

Behavioural signs associated with pain

Pain is highly subjective, and its effects can vary greatly between individuals. Any of the following signs could indicate that a cat is in pain but none of them is necessary to signify the presence of pain. Other factors such as fear, stress and the strength of the cat's motivation to perform an action should also be taken into account because these can also have a very significant influence on an individual's behaviour and reaction to pain.

Reduced activity

Cats in pain may spend more time sleeping and less time exploring, hunting or playing. If they have outdoor access they may spend less time outside. This might be due to physical difficulty in getting through cat flaps or open windows, or because of an increased fear and reduced ability to escape from potential threats that may be encountered outside, such as dogs, predatory wildlife and rival cats.

Reduced mobility

Pain can make it more difficult for a cat to jump up onto furniture or other elevated spaces, causing increased hesitancy when the cat attempts to jump up or resulting in the cat changing its preferred resting or hiding areas to places that are more easily accessible. Pain may also cause a cat to avoid using stairs or use them less often.

House-training issues

A cat in pain might be more likely to toilet indoors if he experiences difficulty in getting outside or is reluctant to go out because of potential threats.

Climbing in and out of a high-sided litter tray, digging into cat litter or walking on the uneven surface of cat litter can also become more difficult, especially for a cat with any form of musculoskeletal pain.

Pain experienced during elimination (urinating or defecating) is another factor to be aware of because the cat can associate the pain with the location, surface or substrate on which it is attempting to eliminate. A learned aversion of the litter tray may then develop, resulting in active litter tray avoidance and inappropriate indoor toileting.

Changes in grooming behaviour

Self-grooming and other maintenance behaviours such as scratching to help condition the claws can decrease due to pain or discomfort, resulting in long brittle nails and poor coat condition. However, over-grooming, especially of one particular area, can also occur and be a sign that the cat may be experiencing pain or discomfort in, or close to, that area.

Temperament changes

Pain can increase irritability and significantly lower the threshold for aggression, especially if being touched causes pain, or if the cat anticipates pain when touched or when about to be touched (Fig. 6.1). Cats in pain might also become generally less tolerant with people and with other household pets and may actively avoid interactions by spending more time hiding or retreating to areas well away from other household members.

Fig. 6.1. A cat may become aggressive if being touched causes or increases pain, or if it anticipates pain when touched or about to be touched.

Disease

A disease is any disorder or abnormal condition of a body part, organ or system. Diseases can have many causes including inflammation, infection (bacterial, viral, fungal or parasitic), trauma, toxins or just simple 'wear and tear'. There are numerous feline diseases, many of which can cause behavioural changes, either directly by affecting the brain and/or neurological system or indirectly by causing other physiological changes that may, in turn, influence the animal's behaviour.

Diseases that directly affect the central nervous system can cause symptoms such as seizures, tremors, disorientation, incoordination, circling/pacing, and/or uncharacteristic mood changes such as unexplained fear, anxiety and/or aggression.

As well as the behavioural changes associated with pain or discomfort as described previously, the physiological effects of disease that might influence behaviour indirectly can include:

- Increased frequency and/or urgency of urination or defecation, potentially leading to house-soiling problems.
- Increased appetite that might lead to increased competition and conflict with other cats or other household pets.
- Pruritus (itching) leading to over-grooming and irritability.
- General malaise and 'feeling unwell' that might lead to irritability and anxiety.

Old Age

As cats increase in age they become more vulnerable to weakness, sensory decline and poor health in general, plus the strength of their immune function can decrease making them increasingly prone to infections. The prevalence of diseases such as hyperthyroidism, renal failure and degenerative joint disease also increases with age, all of which can increase anxiety, irritability and generally reduce the cat's overall coping abilities (see the next section on stress). Behaviour problems linked with pain, discomfort and disease are therefore more likely to be seen in senior cats.

Cognitive dysfunction syndrome (CDS) can also occur in elderly cats (see Box 6.1) and can be responsible for a severe decline in cognitive ability and associated behaviour.

> **Box 6.1. Feline cognitive dysfunction syndrome.**
>
> Cognitive dysfunction syndrome (CDS) is an age-related condition, similar to human dementia, that can affect cats of 10 years of age and older. Studies have shown evidence of physical brain changes including cerebral atrophy and neuronal damage in affected cats.
>
> **Behavioural signs associated with CDS:**
>
> Spatial disorientation – getting trapped or lost in locations well known to the cat. Not being able to find the food bowl or litter tray
> Confusion
> Anxiety
> Aimless wandering or pacing
> Decreased awareness or response to stimuli
> Altered sleep and activity patterns – cats with this condition may be more restless and vocal at night
> Increased and/or inappropriate vocalizations
> Changes in social relationships with owners or other pets, e.g. increased attention seeking or uncharacteristic aggression
> House soiling
> (Gunn-Moore *et al.*, 2007; Landsberg *et al.*, 2010.)

Stress

Stress is a reaction to a physiological, psychological or emotional challenge. A **stressor** is an event, circumstance or stimulus that causes stress. Two types of stress have been described (Selye, 1974, cited in McGowen *et al.*, 2006):

- **Eustress** ('positive stress') is where the challenge is within the individual's ability to cope and the outcome may be beneficial to the individual's state of well-being.
- **Distress** ('negative stress') is where the challenge is beyond the individual's ability cope with an outcome that is damaging to the individual's state of well-being.

It has been more recently proposed, however, that the term 'stress' should only refer to 'distress' or more specifically to environmental challenges that the individual is unable to cope with, especially to situations that he or she is unable to predict or have any control over (Koolhaas *et al.*, 2011).

The physiological stress response

When an animal perceives a potential threat or challenge to its environment, two physiological systems are activated.

The sympathetic–adrenal medullary axis (SAM axis)

This involves the **autonomic nervous system** that is part of the peripheral nervous system responsible for the regulation of bodily functions not under conscious control, such as breathing, heartbeat and digestion. It is made up of two halves: the **sympathetic** and the **parasympathetic**. Under normal circumstances both systems work together to maintain a state of homeostasis but during the stress response sympathetic activity is increased and parasympathetic activity decreased. This takes place via the following process:

- Information received via the senses is sent to the amygdala – a group of neurons located within the limbic system of the brain.
- If a potential threat or challenge is perceived, the amygdala sends a message to the hypothalamus, which in turn sends signals to the adrenal glands – two small endocrine glands situated just above the kidneys.
- Chemicals known as catecholamines, predominantly adrenaline (epinephrine) and noradrenaline (norepinephrine), are then released into the blood stream from the adrenal medulla (the centre of the adrenal gland).
- The effects of adrenaline and noradrenaline are to prepare the animal for 'fight or flight' by activating the following physiological effects:
    - Increased heart rate and blood pressure, which increases blood flow to the muscles.
    - Narrowing of some blood vessels to reduce blood flow to non-essential organs.
    - Widening of air passages to allow increased intake of oxygen.
    - Increased hydrolysis of glycogen to glucose to increase energy availability.

The hypothalamus–pituitary–adrenal axis (HPA axis)

- When a signal of potential threat reaches the hypothalamus it also releases corticotropin releasing hormone (CRH).
- CRH acts on the pituitary gland stimulating increased synthesis and release of adrenocorticotropic hormone (ACTH).
- ACTH is carried to the adrenal glands where it stimulates the release of glucocorticoids (also known as glucocorticosteroids) into the blood stream from the adrenal cortex (the outer part of the adrenal gland).
- Glucocorticoids play an important role in creating energy from the metabolism of carbohydrates, proteins and fats, although they may also have other roles in the stress response that are not as yet fully understood (Sapolsky *et al.*, 2000).
- Once an optimum level in the bloodstream is reached, glucocorticoids are transmitted back to the hypothalamus in a 'negative feedback loop' where they shut down the production of CRH and stop both the SAM and HPA axes.
- However, chronic stress can disrupt this negative feedback cycle, resulting in the continued release of glucocorticoids (Mizoguchi *et al.*, 2003).

The stress response performs an important function in that it increases the individual's ability to escape from, or defend itself against, an actual or potential danger. But if stress is persistent or frequently repeated over an extended period the emotional and physiological changes, especially the continued release of stress hormones, can be directly damaging to both health and welfare.

The impact of stress on feline welfare

The welfare of an animal can be measured by the following factors:

- Physical health.
- The ability and opportunity to perform normal or 'natural' behaviours.
- Mental and emotional health (Casey and Bradshaw, 2007).

Stress can be highly damaging to the welfare of pet cats as it can negatively influence all of these factors.

The Impact of Stress on Physical Health

The urinary system

Feline idiopathic cystitis

Probably the most well-known and well-accepted link between feline stress and disease is the association with feline idiopathic cystitis (FIC).

The term feline lower urinary tract disease (FLUTD) is used to describe a variety of diseases of the bladder and urethra, and is a common and potentially serious health issue in cats. Symptoms of FLUTD can include any or all of the following:

- Difficulty in urinating: the cat may strain to urinate but produce very little or no urine.
- Pain when urinating: the cat may cry out when trying to urinate.
- Increased frequency of urination or attempts to urinate.
- Blood in the urine.
- Urinating in inappropriate and often multiple places. This can be partly due to discomfort causing an 'urgent' need to urinate but can also be due to a learned association with pain and the locations where the cat attempts to urinate.
- Excessive licking of the penis or vulva.
- Over grooming, especially of the lower abdomen and perineum (Bowen and Heath, 2005; Griffiths, 2016).

Causes of FLUTD include:

- Uroliths: bladder stones or stones in the urethra.
- Urinary crystals.
- Trauma.
- Bacterial infection.
- Neoplasia (tumour).

However, the majority of cases (more than 50%) have been found to be idiopathic, meaning that no underlying physical cause can be identified despite affected cats showing signs of bladder inflammation (cystitis), such as blood in the urine, symptoms of discomfort and associated behavioural changes (Buffington *et al.*, 1997; Gerber *et al.*, 2005).

Although the primary cause of FIC is not as yet understood, increasing evidence points to stress as being a major contributory factor. Clinical signs are more often seen during or soon after periods of stress (Jones et al., 1997; Cameron et al., 2004). Affected cats may also show other behavioural signs and physical disorders considered to be stress related. For example:

- Indoor urine spraying.
- Cardiovascular abnormalities.
- Gastrointestinal signs.
- Dermatological signs.
- Exaggerated acoustic startle response (meaning that the cat is more likely to react fearfully to sounds) (Buffington et al., 2006; Stella et al., 2011; Sparkes et al., 2016).

Some cats appear to be predisposed to the condition and may be chronic sufferers or have frequently recurring episodes that can often be self-resolving within a few days (Gunn-Moore, 2003). This can add to the difficulty in diagnosing the disease because by the time the veterinary surgeon examines the cat it may no longer be showing clinical signs.

Exposure to early life stressors, including prenatal stress (see Chapter 5), could be particularly influential in the development of this syndrome (Buffington, 2011; Buffington et al., 2014).

The immune system

It is well established that prolonged production of glucocorticoids and catecholamines as a result of chronic stress can interfere with and suppress the body's immune system. Other, less well understood mechanisms associated with stress might also be involved (Padgett and Glaser, 2003). Being subjected to early life stressors, such as maternal separation or exposure to maternal stress hormones before birth (see Chapter 5), has also been shown to have a negative influence on an individual's lifelong resistance to disease (Avitsur et al., 2006; Mills, 2016). Stress may therefore be a factor in the development and/or spread of many feline infections. There are certain diseases where long-term stress has definitely been shown to be an issue.

Feline infectious peritonitis

Feline coronavirus (FCoV) is the infectious agent that causes feline infectious peritonitis (FIP). It is common in the domestic cat population but not all cats that test positive for FCoV go on to develop FIP. The majority will remain healthy or develop no more than a mild enteritis. Stress has been shown to be the main predisposing factor leading to the development of FIP in cats exposed to FCoV (Addie et al., 2009).

Feline herpesvirus

Feline herpesvirus (FHV & FeHV-1) is a major cause of upper respiratory and ocular disease. Stress may produce a reactivation of symptoms in cats that have previously had the disease, or they may become latent carriers of the infection, becoming a risk to other cats during or following a period of stress (Gaskell *et al.*, 2007).

Upper respiratory tract infection

As well as feline herpesvirus, other diseases including feline calicivirus (FCV), chlamydia, bordetella and mycoplasma have been identified as causes of upper respiratory tract infections in cats (Bannasch and Foley, 2005). Tanaka *et al.* (2012) found that cats admitted to a rescue shelter were more likely to exhibit high levels of stress and were five times more likely to develop an upper respiratory tract infection than those exposed to less stressful situations.

Dermatological conditions

Psychogenic alopecia (stress-related over-grooming)

Short episodes of self-grooming are a normal feline displacement activity and response to acute, mild to moderate stressors (van den Bos, 1998). But excessive grooming resulting in hair thinning or bald patches can occur as a behavioural response to severe chronic stressors. This is often referred to as 'psychogenic alopecia' but it cannot correctly be considered as a 'true alopecia' (hair loss) because, rather than the hair falling out, it is physically pulled out by the cat or bitten off near the root.

Cats can sometimes be quite clandestine when over-grooming so the behaviour might not be obvious to the casual observer. One way to tell if hair thinning or bald patches are due to over-grooming or hair loss is by the feel of the skin in the affected area. If the hair has fallen out the skin will be fairly smooth to the touch. Whereas in cases of over-grooming, the area is more likely to feel rough and 'stubbly' from where the hair has been bitten off close to the root and from hair regrowth (Bowen and Heath, 2005).

Over-grooming can occur anywhere on the body that the cat can reach with its tongue, but it is more common in areas that are easy for the cat to reach, such as on the flanks or abdomen. The cat may also obsessively bite and pull at the feet and claws. Another symptom of over-grooming can be regular regurgitation of hairballs and in some cases this may be the only apparent symptom as the loss of hair might not be sufficient to be visually evident without close examination.

However, stress is far from the only reason for over-grooming and in the majority of cases the primary cause is medical, rather than, or as well as, behavioural (Waisglass *et al.*, 2006). Stress may exacerbate a behaviour initially triggered by a physical cause but over-grooming that is purely behavioural is rare (Hobi *et al.*, 2011). Veterinary investigation, possibly involving a veterinary dermatologist, should therefore always be carried out before assuming the problem to be purely stress related.

Medical differentials for over-grooming include:

- Pathological skin conditions (infectious, allergic and/or parasitic dermatitis) that cause itching or discomfort.
- Internal pain or discomfort, such as that caused by feline lower urinary tract disease (Fig. 6.2) (Bowen and Heath, 2005; Griffiths, 2016).
- Neurological conditions.

Fig. 6.2. In most cases of over-grooming the primary cause of the behaviour is medical but this can be exacerbated or maintained by stress. Over-grooming of the abdomen may be indicative of internal discomfort such as that caused by a lower urinary tract infection. Photo courtesy of Celia Haddon.

Pathological skin diseases

Human and canine studies have demonstrated an association between stress and pruritic or inflammatory skin diseases (Kimyai-Asadi and Usman, 2001; Nagata *et al.*, 2002). Further research is required to ascertain if a similar link exists in cats but it is very possible that it does, especially in cases of allergic skin diseases such as flea allergic dermatitis (Sparkes *et al.*, 2016).

The gastrointestinal system

There is a direct link between the gut and the brain via the sympathetic and parasympathetic pathways (Bhatia and Tandon, 2005). An association between stress and gastrointestinal disease is therefore as likely in cats as it is in other species.

Irritable bowel disease

In human and canine medicine there is a well-recognized link between psychological stress and irritable bowel disease (IBD) (Simpson, 1998). There is less scientific evidence that the condition exists in cats but it is generally believed that it does, and as with other species it is likely that there is an association with chronic stress (Sparkes *et al.*, 2016).

Alterations in gastrointestinal motility

Stress has been shown to both delay gastric emptying and accelerate large bowel motility (Taché *et al.*, 2001; Bhatia and Tandon, 2005). It can therefore have varying effects of reduced appetite, vomiting and diarrhoea. Stella *et al.* (2011) found 'sickness behaviours', most notably vomiting or regurgitation of food, bile or hair (hairballs), diarrhoea or constipation, to be a consequence of cats subjected to periods of stress.

Reduced appetite and water intake, plus a reduction in the frequency of elimination leading to constipation, is a well-known and common response by cats to periods of stress associated with hospitalization, cattery visits, re-homing, etc. If food intake is severely reduced this can even result in the potentially fatal condition of hepatic lipidosis.

A reduction in upper gastrointestinal motility caused by chronic stress might also explain why some cats vomit or regurgitate hairballs on a regular basis (Sparkes *et al.*, 2016).

Obesity

Raised glucocorticoid levels have been shown to increase appetite in humans (Tataranni *et al.*, 1996) and overeating is a well-recorded reaction to chronic stress in both humans and animals (McMillan, 2013). Obesity in cats may therefore also be linked to chronic environmental stress. This might explain why standard weight loss programmes for cats can sometimes be ineffective if underlying causes of stress are not also identified and addressed (German and Heath, 2016).

Enforced diets that reduce the amount of food offered can also increase stress, not only because of hunger and the reduction of food available, but in multi-cat households reduced resource availability can be a common cause of tension and conflict.

Perceived competition with other cats can also increase the amount of food that a cat may eat at one time and the rate of food intake, causing the cat to eat much faster than it would do normally. This may also contribute to overeating and obesity because the cat might be more likely to demand food earlier than it would do if eating little and often. However, another effect of eating too much too fast can be vomiting or regurgitation of food soon after eating, which if happening frequently can result in weight loss and poor condition.

The endocrine system

Diabetes mellitus

Diabetes mellitus is a metabolic condition in which the body is unable to utilize the energy in food, leading to high blood-sugar levels and potentially serious health complications.

The body's blood-sugar level is regulated by the hormone insulin, which enables the cells to utilize sugar (glucose) as energy and helps the liver to store excess glucose, releasing it as required.

There are two types of diabetes mellitus:

- Type 1: the inability to produce sufficient insulin.
- Type 2: resistance to insulin.

Chronic stress and obesity are recognized as important contributory factors in the development of type 2 diabetes in humans (Spruijt-Metz et al., 2014). Although there is a lack of specific studies on cats with the condition, it is very possible that there could also be a causal link between stress and type 2 diabetes in cats (Sparkes et al., 2016).

Hyperthyroidism

Hyperthyroidism (excessive production of thyroid hormone) is a common condition in middle-aged and older cats. Cats with hyperthyroidism have been found to have more cortisol (one of the corticosteroids produced as part of the stress response) than creatinine (a normal metabolic breakdown product) in their urine, indicating a higher than normal level of plasma cortisol and a possible link between the disease and chronic stress (de Lange et al., 2004). It is unclear, however, if this is cause or effect because the disease is also known to activate the stress response.

The neurological system

Feline oral facial pain syndrome (FOPS)

This is a particularly unpleasant condition causing cats to suffer from abnormally severe neuropathic pain of the mouth and facial area. Cats with the condition claw intensely at the mouth causing lacerations to the face and tongue.

The condition is most common in Burmese, suggesting a possible hereditary susceptibility, but any breed can be affected. Predisposing factors can be anything that causes pain or discomfort in the mouth such as, eruption of permanent teeth, dental disease, or mouth ulceration. But the apparent pain and the cat's reactive behaviour is far greater than would normally be expected. Environmental stress such as conflict with other cats, visits to catteries or the veterinary clinic can be trigger factors, and chronic stress may also play a role in the development of the disease as it appears that cats with poor coping strategies in multi-cat households are more likely to be predisposed to the condition (Rusbridge et al., 2010).

Feline hyperaesthesia syndrome

Hyperaesthesia is an abnormal sensitivity and reaction to touch or movement of the skin. It is often episodic and may be triggered by touch, either by self-grooming, being petted or increased arousal, for example during play.

Hyperaesthesia is characterized by symptoms including:

- Rippling or twitching of the skin.
- Acute biting, scratching or licking, especially in the lumbar area near the tail base.
- Erratic, agitated or sometimes aggressive behaviour, which can sometimes appear to be self-directed.

As with FOPS the exact cause of hyperaesthesia is unknown. It is likely that it is not a separate disease but a collection of symptoms related to a wide range of undiagnosed, skin, musculoskeletal or nervous system conditions. Exposure to long-term stressors appears to be a predisposing factor, and acute stressors may act as a trigger factor (Bowen and Heath, 2005; Heath and Rusbridge, 2007).

The cardiovascular system

Because of their direct influence on the heart and blood pressure, prolonged or frequent release of catecholamines can have potentially damaging effects on the cardiovascular system (Cohen *et al.*, 2007).

Lifespan

Human and canine studies have shown that chronic stress can shorten lifespan and accelerate ageing by causing changes to DNA (Dreschel, 2010; Ahola *et al.*, 2012). Although there are no specific studies on this in cats, it is possible that they may suffer the same effect.

The Impact of Stress on Mental and Emotional Health

Emotions such as fear, anxiety and frustration are all a part of stress, which can be detrimental to mental health and welfare.

Chronic anxiety can increase the risk of mental health issues in humans and it is possible that there is a similar effect in animals. For example, there is a recognized link between anxiety and compulsive disorders in animals (Overall, 1998). Stereotypic behaviours in animals can also arise from long-term anxiety, conflict or frustration (Mason, 1991). A further effect of stress is an increase in the rate of cognitive decline in animals already suffering from cognitive dysfunction syndrome (Mills *et al.*, 2014).

Stress and common feline behaviour problems

Aggression

Sympathetic activation produces the 'flight or fight' response and increased startle and defensive reactions. If a cat is restrained, otherwise unable to escape or too close to a perceived threat that an attempt to escape could place the cat at increased risk, then aggression can be the only remaining means of defence.

Persistent 'background' stress can also raise long-term arousal sufficiently to produce exaggerated responses, e.g. startle or aggression in reaction to minor or moderate threats or stressors that might otherwise produce very little or no reaction.

Re-directed aggression

If a cat is unable to flee from, chase away, or fight with a perceived threat, the heighten sense of arousal and frustration it experiences may cause it to re-direct aggression towards whatever or whoever is closest. This could be the owner or another animal that it normally considers to be part of its social group. Re-directed aggression is a common cause of relationship breakdown and fighting between cats that have previously had a friendly, bonded relationship.

House-soiling

A breakdown in house training might result from the direct effect of stress on the gastrointestinal or urinary systems (see previous sections), contributing to diseases that can cause frequent elimination in multiple locations owing to urgency or litter-box aversion due to associations with pain or discomfort. A stressed, fearful or anxious cat may also want to avoid eliminating in places where it feels vulnerable, which may include the litter box and/or outdoor elimination areas.

Indoor urine spraying

Urine marking (spraying) is a normal communication behaviour in entire cats (male and female) used primarily to advertise fitness and sexual status to potential sexual partners and rivals (see Chapter 3). But in neutered cats, it appears to be linked to a heightened state of arousal due to territorial behaviour and/or stress and anxiety (Borchelt and Voith, 1996; Bowen and Heath, 2005; Neilson, 2009; Amat *et al.*, 2015).

Indoor urine spraying by neutered pet cats seems to occur most commonly when there is perceived competition for resources with other cats, especially with others not considered to be of the same social group. It can also occur in reaction to other household stressors such as: moving house, visitors, building works or a new baby in the household (Borchelt and Voith, 1996; Pryor *et al.*, 2001).

The reasons why neutered cats urine mark in the home is not fully understood and various theories have been suggested, including:

- Territorial behaviour: advertising presence and fitness to keep other cats away and so avoid contact and conflict. However, there is no evidence that cats avoid the spray marks of others. In fact, they are more likely to spend time investigating the mark and so stay in the area longer than they would do normally (Bradshaw and Cameron-Beaumont, 2000).
- A coping strategy: (i) helping the cat to reduce its stress by increasing its scent profile (Bowen and Heath, 2005); or (ii) providing itself with information as to where there is likely to be potential danger or where a perception of threat has been experienced previously, thereby giving itself greater predictability and control over its environment (Bradshaw *et al.*, 2012).

Compulsive behaviours

Over-grooming/self trauma

As mentioned previously over-grooming can be a response to stress in the cat; however, in the majority of cases an underlying medical cause is the more likely reason for the behaviour.

Pica/wool sucking

Pica is the ingestion of inedible items. It is a behaviour that may start with the cat or kitten sucking wool or other similar fabrics and then progressing to ingestion of the original material and/or other items. It has been linked with both early weaning and other potential stressors such as changes associated with life stages and poor environmental enrichment (Bradshaw *et al.*, 1997). However, a more recent study found no association with early weaning or insufficient environmental enrichment but did find that cats exhibiting pica are less likely to be fed ad-lib (Demontigny-Bédard *et al.*, 2016).

Possible medical causes of this behaviour include:

- Diseases affecting the central nervous system, e.g. FeLV associated myelopathy, feline infectious peritonitis and cognitive dysfunction.
- Gastric disorders, e.g. gastric motility disorder, gastric discomfort and polyphagia.
- Oral pain or discomfort (Bowen and Heath, 2005; Frank, 2014).

Acute Stress – the Influence on Physiological Parameters

Stress that is of short duration is less likely to be directly damaging to health and well-being unless frequently and persistently repeated. But acute stress can have short-term effects on physiological parameters including:

- Heart rate.
- Blood pressure.
- Respiratory rate.
- Chemical concentrations in the blood, e.g. glucose.

Because they are short lived, these changes are of little concern but they can provide false results to veterinary tests and examinations and thereby hinder correct diagnoses of disease.

Assessing Stress

Stress in cats can be difficult to assess because every individual will have different reactions to potential stressors and different perceptions of what is considered stressful (Casey and Bradshaw, 2007). Acute stress can be easier to identify than chronic stress but, even then, some cats will demonstrate clear physiological and behavioural signs, whilst others, despite being no less stressed, will show very little outward evidence.

The presence of disease conditions known to be associated with stress is also not a reliable indicator that stress precipitated the disease. Not only are there other contributory causes to these conditions, but also being unwell or in pain and discomfort can in itself be a significant stressor.

Measurements of hormones released into the bloodstream during the stress response, most notably cortisol, one of the major glucocorticoid hormones, can be used in controlled research situations to identify stress in groups of animals. But because of difficulties in correctly interpreting results owing to individual variation in physiological response, plus normal variations owing to factors such as age, sex and even time of day, such tests are of very limited use in assessing stress in individuals. Another problem is that methods used to obtain samples may in themselves involve some degree of stress and so influence results (Casey and Bradshaw, 2007). An additional limitation is that any form of arousal, negative or positive, can raise blood cortisol levels, which can make it a less reliable indicator of distress (Mills *et al.*, 2014).

Observable Signs of Stress in Cats

Physiological signs associated with sympathetic activation

- Panting.
- Salivation.
- Dilated pupils.
- Urination.
- Defecation.
- Anorexia.
- Piloerection.

Behavioural signs of emotional stress

The number and severity of signs exhibited will vary between individuals and in chronic stress they are likely to be seen intermittently:

- Increased vigilance and sleep inhibition, especially when in the presence of a stressor.
- Increased muscle tension, in preparation for 'fight or flight'.
- Lowered head and body posture. The head may be positioned lower than the body.
- Feet remain in contact with the ground when the cat is resting, allowing for a more rapid escape when necessary.
- Ears flattened sideways.
- Tail held close to body.
- Inhibition of normal behaviours - such as eating, grooming, elimination, exploratory behaviour and play (Rochlitz, 1999; Casey and Bradshaw, 2007).

- Avoidance behaviours:
    - Hiding: acute stress can cause short episodes of hiding from stressors such as unusual very loud noises, etc. but a chronically highly stressed house cat may spend most of its time hiding.
    - Fleeing: as with hiding, most cats will run from the things that frightened it; however, a cat subjected to chronic stress is more likely to experience increased arousal and exhibit more frequent or exaggerated responses, and therefore be more likely to try to escape from an increased number of different stimuli.
    - Feigned, sham or defensive sleeping: pretending to sleep or 'forced' unrelaxed sleep is a strategy that some cats use to avoid interactions with perceived or anticipated stressors (Kessler and Turner, 1997; Rochlitz et al., 1998; Casey and Bradshaw, 2007).

References

Addie, D., Belak, S., Boucraut-Baralon, C. et al. (2009) Feline infectious peritonitis. ABCD guidelines on prevention and management. *Journal of Feline Medicine and Surgery* 11, 594–604.

Ahola, K., Sirén, I., Kivimäki, M., Ripatti, S., Aromaa, A., Lönnqvist, J. and Hovatta, I. (2012) Work-related exhaustion and telomere length: a population-based study. *Plos One* 7 (7) e40186. DOI:10.1371/journal.pone.0040186.

Amat, M., Camps, T. and Manteca, X. (2015) Stress in owned cats; behavioural changes and welfare implications. *Journal of Feline Medicine and Surgery* 1–10.

Avitsur, R., Hunzeker, J. and Sheridan, J.F. (2006) Role of early stress in the individual differences in host response to viral infection. *Brain, Behaviour and Immunity* 20, 339–348.

Bannasch, M.J. and Foley, J.E. (2005) Epidemiologic evaluation of multiple respiratory pathogens in cats in animal shelters. *Journal of Feline Medicine and Surgery* 7, 109–119.

Bhatia, V. and Tandon, R.K. (2005) Stress and the gastrointestinal tract. *Journal of Gastroenterology and Hepatology* 20, 332–339.

Borchelt, P.L. and Voith, V.L. (1996) Elimination behavior problems in cats. In: Voith, V.L. and Borchelt, P.L. (eds) *Readings in Companion Animal Behavior*. Veterinary Learning Systems, Trenton, New Jersey.

Bowen, J. and Heath, S. (2005) *Behaviour Problems in Small Animals: Practical Advice for the Veterinary Team*. Saunders Ltd, Elsevier, Philadelphia, Pennsylvania, USA.

Bradshaw, J.W.S. and Cameron-Beaumont, C. (2000) The signalling repertoire of the domestic cat and its undomesticated relatives. In: Turner, D.C. and Bateson, P. (eds) *The Domestic Cat: The Biology of its Behaviour*. Cambridge University Press, Cambridge, UK.

Bradshaw, J.W.S., Neville, P.F. and Sawyer, D. (1997) Factors affecting pica in the domestic cat. *Applied Animal Behaviour Science* 52, 373–379.

Bradshaw, J.W.S., Casey, R.A. and Brown, S.L. (2012) *The Behaviour of the Domestic Cat*, 2nd edn. CAB International, Wallingford, UK.

Buffington, C.A.T. (2011) Idiopathic cystitis in domestic cats – beyond the lower urinary tract. *Journal of Veterinary Internal Medicine* 25, 784–796.

Buffington, C.A.T., Chew, D.J., Kendall, M.S., Scrivani, P.V., Thompson, S.B., Blaisdell, J.L. and Woodworth, B.E. (1997) Clinical evaluation of cats with non-obstructive urinary tract disease. *Journal of the American Veterinary Medical Association* 210, 46–50.

Buffington, C.A.T., Westropp, J.L., Chew, D.J. and Bolus, R.R. (2006) Risk factors associated with clinical signs of lower urinary tract disease in indoor-housed cats. *Journal of the American Veterinary Medical Association* 228, 722–725.

Buffington, C.A.T., Westropp, J.L. and Chew, D.J. (2014) From FUS to pandora syndrome. Where are we, how did we get here and where to now? *Journal of Feline Medicine and Surgery* 16, 385–394.

Cameron, M.E., Casey, R.A., Bradshaw, J.W.S., Waran, N.K. and Gunn-Moore, D.A. (2004) A Study of environmental and behavioural factors that may be associated with feline idiopathic cystitis. *Journal of Small Animal Practice* 45, 144–147.

Casey, R.A. and Bradshaw, J.W.S (2007) The assessment of welfare. In: Rochlitz, I. (ed.) *The Welfare of Cats*. Springer, Dordrecht, The Netherlands.

Cohen, S., Janicki-Deverts, D. and Miller G.E. (2007) Psychological stress and disease. *Journal of the American Medical Association* 298, 1685–1687.

De Lange, M.S., Galac, S., Trip, M.R.J. and Kooistra, H.S. (2004) High urinary corticoid/creatinine ratios in cats with hyperthyroidism. *Journal of Veterinary Internal Medicine* 18, 152–155.

Demontigny-Bédard, I., Beauchamp, G., Bélanger, M.C. and Frank, D. (2016) Characterization of pica and chewing behaviors in privately owned cats: a case-control study. *Journal of Feline Medicine and Surgery* 18, 652–657.

Dreschel, N.A. (2010) The effects of fear and anxiety on health and lifespan in pet dogs. *Applied Animal Behaviour Science* 125, 157–162.

Frank, D. (2014) Recognizing behavioral signs of pain and disease: a guide for practitioners. *Veterinary Clinics of North America: Small Animal Practice* 44, 507–524.

Gaskell, R., Dawson, S., Radford, A. and Thiry, E. (2007) Feline herpesvirus. *Veterinary Research* 38, 337–354.

Gerber, B., Boretti, F.S., Kley, S., Laluha, P., Müller, C. et al. (2005) Evaluation of clinical signs and causes of lower urinary tract disease in European cats. *Journal of Small Animal Practice* 46, 571–577.

German, A. and Heath, S. (2016) Feline obesity. In: Rodan, I. and Heath, S. (eds) *Feline Behavioural Health and Welfare*. Elsevier, St Louis, Missouri, USA, pp. 148–161.

Griffiths, K. (2016) How I approach overgrooming in cats. *Veterinary Focus* 26, 32–39.

Gunn-Moore, D.A. (2003) Feline lower urinary tract disease. *Journal of Feline Medicine and Surgery* 5, 133–138.

Gunn-Moore, D., Moffat, K., Christie, L.A. and Head, E. (2007) Cognitive dysfunction and the neurobiology of aging in cats. *Journal of Small Animal Practice* 48, 546–553.

Hardie, E.M., Roe, S.C. and Martin, F.R. (2002) Radiographic evidence of degenerative joint disease in geriatric cats: 100 cases (1994–1997). *Journal of the American Veterinary Medical Association* 220, 628–632.

Heath, S. and Rusbridge, C. (2007) Feline hyperaesthesia syndrome and orofacial pain in Burmese. *Scientific Proceedings of ESFM Feline Congress*. Prague, pp. 79–82.

Hobi, S., Linek, M., Marignac, G., Olivrys̨, T., Beco, L. et al. (2011) Clinical characteristics and causes of pruritus in cats: a multicentre study on feline hypersensitivity – associated dermatoses. *Veterinary Dermatology* 22, 406–413.

Jones, B.R., Samson, R.L. and Morris, R.S. (1997) Elucidating the risk factors of feline lower urinary tract disease. *New Zealand Veterinary Journal* 45, 100–108.

Kessler, M.R. and Turner, D.C. (1997) Stress and adaptation of cats (*Felis Silvestris Catus*) housed singly, in pairs and in groups in boarding catteries. *Animal Welfare* 6, 243–254.

Kimyai-Asadi, A. and Usman, A. (2001) The role of psychological stress in skin disease. *Journal of Cutaneous Medicine and Surgery* 5, 140–145.

Koolhaas, J.M., Bartolomucci, A., Buwalda, B. et al. (2011) Stress revisited: a critical evaluation of the stress concept. *Neuroscience and Biobehavioral Reviews* 35, 1291–1301.

Landsberg, G., Denenberg, S. and Araujo, J. (2010) Cognitive dysfunction in cats. A syndrome we used to dismiss as 'old age'. *Journal of Feline Medicine and Surgery* 12, 837–848.

Lascelles, B.D.X, Henry III, J.B., Brown, J., Robertson, I., Sumrell, A.T. et al. (2010) Cross-sectional study of the prevalence of radiographic degenerative joint disease in domesticated cats. *Veterinary Surgery* 39, 535–544.

Mason, G.J. (1991) Stereotypies: a critical review. *Animal Behaviour* 41, 1015–1037.

McMillan, F.D. (2013) Stress-induced and emotional eating in animals: a review of the experimental evidence and implications for companion animal obesity. *Journal of Veterinary Behaviour* 8, 376–385.

Mills, D. (2016) What are stress and distress, and what emotions are involved? In: Sparkes, A. and Ellis, S. (eds) *ISFM Guide to Feline Stress and Health: Managing Negative Emotions to Improve Feline Health and Wellbeing*. International Cat Care, Tisbury, Wiltshire, UK, pp. 7–18.

Mills, D., Karagiannis, C. and Zulch, H. (2014) Stress – its effects on health and behaviour: a guide for practitioners. *Veterinary Clinics of North America: Small Animal Practice* 44(3), 525–541.

Mizoguchi, K., Ishige, A., Aburada, M. and Tabira, T. (2003) Chronic stress attenuates glucocorticoid negative feedback involvement of the prefrontal cortex and hippocampus. *Neuroscience* 119(3), 887–897.

Nagata, M., Shibata, K., Irimajiri, M. and Luescher, A.U. (2002) Importance of psychogenic dermatoses in dogs with pruritic behaviour. *Veterinary Dermatology* 13(4), 211–229.

Neilson, J.C. (2009) House soiling by cats. In: D.F. Horwitz and D.S. Mills (eds) *BSAVA Manual of Canine and Feline Behavioural Medicine*, 2nd edn. BSAVA, Gloucester, UK.

Overall, K. (1998) Self-injurious behaviour and obsessive-compulsive disorder in domestic animals. In: Dodman, N.H. and Shuster, L. (eds) *Psychopharmacology of Animal Behavior Disorders*. Blackwell Science, Malden, Massachusetts, USA, pp. 222–252.

Padgett, D.A. and Glaser, R. (2003) How stress influences the immune response. *Trends in Immunology* 24, 444–448.

Pryor, P.A., Hart, B.L., Bain, M.J. and Cliff, K.D. (2001) Causes of urine marking in cats and effects of environmental management on frequency of marking. *Journal of the American Veterinary Medical Association* 12, 1709–1713.

Rochlitz, I. (1999) Recommendations for the housing of cats in the home, in catteries and animal shelters, in laboratories and in veterinary surgeries. *Journal of Feline Medicine and Surgery* 1, 181–191.

Rochlitz, I., Podberscek, A.L. and Broom, D.M. (1998) Welfare of cats in a quarantine cattery. *The Veterinary Record* 143, 35–39.

Rusbridge, C., Heath, S., Gunn-Moore, D.A., Knowler, S.P., Johnston, N. and McFadyen, A.K. (2010) Feline orofacial pain syndrome (FOPS): a retrospective study of 113 cases. *Journal of Feline Medicine and Surgery* 12(6), 498–508.

Sapolsky, R.M., Romero, L.M. and Munck, A.U. (2000) How do glucocorticoids influence stress responses? Integrating permissive, suppressive, stimulatory and preparative actions. *Endocrine Review* 21, 55–89.

Selye, H. (1974) *Stress Without Distress*. J.B. Lippincott, New York. Cited in: McGowen, J., Gardner, D. and Fletcher, R. (2006) Positive and negative affective outcomes of occupational stress. *New Zealand Journal of Psychology* 35(2), 92–98.

Simpson, J.W. (1998) Diet and large intestinal disease in dogs and cats. *The Journal of Nutrition* 128, 2717S–2722S.

Sparkes, A., Bond, R., Buffington, T., Caney, S., German, A., Griffin, B. and Rodan, I. (2016) Impact of stress and distress on physiology and clinical disease in cats. In: Sparkes, A. and Ellis, S. (eds) *ISFM Guide to Feline Stress and Health: Managing Negative Emotions to Improve Feline Health and Wellbeing*. International Cat Care, Tisbury, Wiltshire, UK, pp. 41–53.

Spruijt-Metz, D., O'Reilly, G.A., Cook, L., Page, K.A. and Quinn, C. (2014) Behavioral contributions to the pathogenesis of type 2 diabetes. *Current Diabetes Reports* 14, 1–10.

Stella, J.L., Lord, L.K. and Buffington, C.A.T. (2011) Sickness behaviors in response to unusual external events in healthy cats and cats with feline interstitial cystitis. *Journal of the American Veterinary Medical Association* 238, 67–73.

Taché, Y., Martinez, V., Million, M. and Wang, L. (2001) Stress and the gastrointestinal tract III. Stress-related alterations of gut motor function: role of brain corticotropin-releasing factor receptors. *American Journal of Physiology – Gastrointestinal and Liver Physiology* 280, G173–G177.

Tanaka, A., Wagner, D.C. and Kass, P.H. (2012) Associations with weight loss, stress and upper respiratory tract infection in shelter cats. *Journal of the American Veterinary Medical Association* 240, 570–576.

Tataranni, P.A., Larson, D.E., Snitker, S., Young, J.B., Flatt, J.P. and Ravussin, E. (1996) Effects of glucocorticoids on energy metabolism and food intake in humans. *American Journal of Physiology – Endocrinology and Metabolism* 271, E317–E325.

Van den Bos, R. (1998) Post-conflict stress-response in confined group-living cats (*Felis silvestris catus*). *Applied Animal Behavioural Science* 59, 323–330.

Waisglass, S.E., Landsberg, G.M., Yager, J.A. and Hall, J.A. (2006) Underlying medical conditions in cats with presumptive psychogenic alopecia. *Journal of the American Veterinary Medical Association* 228, 1705–1709.

7 Learning, Training and Behaviour

Not all feline behaviour is innate or governed purely by instinct, much of an individual cat's behaviour develops via learning. So, to better understand feline behaviour it is necessary to understand how cats learn and how what they learn can influence their actions and their relationship with us.

There is a common misconception that cats cannot be trained, but any animal capable of learning is also capable of being trained. How animals learn is very much the same regardless of species; however, what they are capable of learning and the most effective ways of training can vary greatly because of species-specific differences in physiology, cognition and 'natural' behaviour. There are without doubt some similarities in dog and cat training but if we attempt to train a cat in exactly the same way as we would a dog and expect a cat to behave in the same way as a dog, we are much less likely to be successful than if we tailor our expectations and training methods especially for cats.

Why Train Cats?

Training can help in many ways to improve the cat–owner relationship (Fig. 7.1). Not only can it provide owners with some element of control over their pet's behaviour and help to prevent some behaviour problems, it can also improve the general welfare of pet cats by increasing mental enrichment and stimulation and help them to cope better with life as a pet in a human household (Bradshaw and Ellis, 2016). The treatment of behaviour problems can also involve some training or manipulation of learning.

Learning Theory

Learning theory is the overall term used to describe the various ways in which learning takes place.

Habituation

Fear and defensive behaviour is a normal response to novel or unidentified stimuli, but if an animal were to continue to react fearfully to frequently occurring stimuli that prove to be harmless the result would be sensory overload, leading to increased stress and reduced welfare. Habituation, one of the simplest forms of learning, is the process whereby the animal becomes accustomed to and learns to ignore everyday non-threatening sights, sounds and smells.

Fig. 7.1. Training can help to improve feline welfare and enhance the cat–owner relationship. Photo courtesy of Celia Haddon.

Habituation is more likely to occur if the stimulus is repeated frequently within a relatively short period and is at a level that is within the individual's ability to cope. It also occurs more readily and easily during a kitten's sensitive period of development, between 2 and 7 weeks of age (see Chapter 5), after which time there can be an increased risk of sensitization rather than habituation.

Practical feline example

- Common household sounds, such as those produced by a washing machine or television may be considered potentially frightening and produce a startle response and fear reaction if a cat or kitten has not previously encountered them. But with repeated exposure at a relatively low level, the cat or kitten can become 'habituated' to these sounds and learn to ignore them, even if they are later presented at an increased level, for example if the volume of the television is turned up.

Sensitization

Sensitization can be considered the opposite of habituation in that repetition causes the animal to become more rather than less reactive to a stimulus. Sensitization is more likely to occur if the stimulus is unpredictable and presented at a level that is beyond the individual's ability to cope. The risk of sensitization is also greater the more emotionally aroused, e.g. fearful, an animal is when the stimulus is presented (Davis, 1974).

Practical feline examples

- It is very common for both dogs and cats to become sensitized and highly fearful of fireworks. This is because fireworks are very loud and unpredictable. Plus, the sound of the first firework going off is likely to increase the animal's fear arousal sufficiently to allow sensitization to occur more readily with each subsequent repetition.
- Cats are also more likely to become sensitized to potentially frightening sights, sounds and even smells when at the veterinary practice or when travelling, owing to an already heightened state of fear arousal at the time.

Associative Learning

Habituation and sensitization are the most basic forms of learning that stem from the presentation of a single stimulus. Associative learning involves two events occurring together and a learned association being made between the two.

Classical conditioning

This is the learned association between a reflex reaction and an unrelated stimulus that would not otherwise cause the reaction.

Reflex reactions

These are reactions to stimuli that occur without any conscious thought or prior experience. For example, touching something hot or otherwise painful can cause a reflexive withdrawal; the smell or taste of food can cause the reflexive production of saliva; and a sudden loud and unexpected sound can produce a reflexive startle response.

Pavlov's conditioned reflexes

Ivan Petrovich Pavlov was a Russian physiologist with an interest in the physiology of digestion. One of his research experiments was designed to measure the salivary secretions of dogs. During his study he noticed that the dogs did not only salivate when food was placed in their mouths but saliva was also produced at other times. He observed that the saliva production was not random but occurred in response to specific stimuli such as the sight or sound of the person that fed the dogs. After becoming aware of this phenomenon, Pavlov altered his experiments in order to discover more about it.

This involved the following process:

1. The presentation of food, with no other stimulus. This caused the production of saliva. Pavlov called this the '*Unconditioned Response*' (UR).

2. The presentation of a sound. This produced no reaction from the dogs. Pavlov called this the '*Unconditioned Stimulus*' (US).
3. The presentation of the same sound, followed 2 seconds later by food.

After a few pairings the sound alone resulted in the production of saliva. The sound was presented for 30 seconds, allowing the experimenter to measure the production of saliva in response to the sound without the presence of food. Pavlov called the production of saliva caused by the sound alone the '*Conditioned Response*' (CR). The sound that then caused the production of saliva he called the '*Conditioned Stimulus*' (CS) (Pavlov, 1927, cited in Lieberman, 1993).

Although the production of saliva is a purely physiological response, cognitive learning, i.e. learning that involves conscious thought, would also have occurred, in that the dogs would have learned to expect food when they heard the sound.

Classical conditioning can apply not only to the anticipation of food, but also to any reflexive or emotional response, including relaxation and fear, which in turn can influence behavioural responses.

Practical feline examples

- The cat runs towards the kitchen when he hears a certain cupboard door opening.
  Enjoyment (and production of saliva) from having a food treat = *Unconditioned Response*.
  The sound of a cupboard opening producing no response = *Unconditioned Stimulus*.
  The sound of the cupboard opening after the cat has associated the sound with being given treats = *Conditioned Stimulus*.
  Anticipation of enjoyment (and production of saliva) when hearing the cupboard opening = *Conditioned Response*.
- The cat runs away when he sees the cat carrier.
  Fear produced by travel, especially trips to the vet = *Unconditioned Response*.
  The sight of the cat carrier producing no response = *Unconditioned Stimulus*.
  The sight of the cat carrier after being taken to the vets a few times = *Conditioned Stimulus*.
  The cat feels frightened as soon as he sees the cat carrier = *Conditioned Response*.
- The cat jumps up on the owner's lap as soon as the TV is turned on.
  Relaxation and feeling good from being petted and being safe, warm and comfortable = *Unconditioned Response*.
  The TV being switched on producing no response = *Unconditioned Stimulus*.
  The TV being switched on after becoming associated with the owner sitting down, relaxing and petting the cat = *Conditioned Stimulus*.
  Anticipation of relaxation and pleasurable interaction with the owner when the TV is switched on = *Conditioned Response*.

Operant conditioning (or instrumental learning)

With classical conditioning, the cat learns to make associations between good or bad feelings and stimuli or events that are out of his control. With operant conditioning the outcome, good or bad, becomes associated with the cat's own actions.

Reinforcement

Reinforcement has occurred if the outcome of a behaviour is rewarding for the animal, resulting in an increased probability that the behaviour will be repeated (Lieberman, 1993).

Positive reinforcement

Positive reinforcement is the direct application of reward following a behaviour, effectively increasing the chances that the behaviour will be repeated.

Practical feline examples

- The cat jumps up on the owner's lap. The owner strokes the cat. The behaviour of jumping up on the owner's lap is rewarded by the owner stroking the cat and the chance that the behaviour will be repeated is increased.
- The cat opens the cupboard door where the treats are kept. The behaviour of pawing at the cupboard door to open it is self-rewarded by the cat being able to get to the treats and the chance that the behaviour will be repeated is increased.

Negative reinforcement

Negative reinforcement is when a behaviour results in relief from an unpleasant experience, increasing the chances that the behaviour will be repeated.

Practical feline examples

- The cat is outside in the cold and rain. He meows loudly. The owner opens the door to let the cat in. The behaviour of meowing at the door is rewarded by the owner opening the door and allowing the cat to get away from the cold and wet. The chance that the behaviour of meowing at the door to be let in is increased.
- The cat is being chased by a dog. The cat jumps up and over a high wall. The dog is unable to follow. The behaviour of jumping over the wall is self-rewarded by the cat getting away from danger. The chance that the cat will jump over the same or a similar high wall when feeling threatened is increased.

Training reward or reinforcer

A reward is something beneficial that results from an action. A reinforcer is something that increases the probability that a behaviour or action will be repeated. The two are often indistinguishable but it cannot be guaranteed that an intended reward will act as a reinforcer. To increase the prospect that a reward will also be a reinforcer it should meet the following criteria:

- It needs to be given as soon as possible after the desired behaviour. The bigger the delay the less chance that the reward will become associated with the behaviour.
- It needs to be sufficiently rewarding (Box 7.1).
- It should be easy to give and receive.
- It should not be so excessively rewarding or overexciting that it increases arousal to a level that disrupts the cat's ability to learn, or results in frustration when the cat is denied access to the reward until it has performed the desired behaviour. During training, the cat should remain calm, interested and keen to receive the reward but should not appear 'obsessed' with it.
- The long-term effectiveness of a reward as a reinforcer can also be influenced by how often it is presented (Box 7.2).

Unintentional reinforcement

One reason why some unwanted learned behaviours appear so difficult to break is that they can easily be unintentionally rewarded; if this occurs intermittently the behaviour becomes more resistant to extinction (see Non-reward, page 109).

Practical feline example

- A cat may learn that pushing items off a shelf is an effective way of getting his owner's attention. The owner tries to ignore the behaviour so that the cat no longer gets the reward of the owner's attention. But when the cat is about to push something delicate and valuable off the shelf the owner still responds, giving the cat the attention it wants, and thereby intermittently reinforcing the behaviour and making the behaviour even more likely to continue.

Self-rewarding behaviour

If a reward occurs following a behaviour without intervention from another individual we can say that the behaviour is 'self-rewarding'. For example, for a cat jumping up onto a work surface and eating any food that has been left there can be a self-rewarding behaviour.

Secondary reinforcement

A **Primary reinforcer** is something that an animal desires and is usually something necessary for survival and/or well-being, for example, food, safety, shelter, comfort or social interaction.

A **Secondary reinforcer** (also known as a **conditioned reinforcer** – see Classical conditioning) is something that has become strongly associated with a primary reinforcer.

The most common secondary reinforcer for humans is money, which is nothing more than paper, metal, plastic or just figures on a computer screen, none of which is

> **Box 7.1. Training rewards: what works best for cats.**
>
> **Food treats**
>
> Food treats may not be sufficiently rewarding if the cat is not food orientated, not hungry at the time of training or very limited in its food preferences.
>
> If the cat does not appear interested in a food reward:
>
> - Try fresh cooked meat, cheese, or fish rather than dry kibble or dried treats (paste or wet food can also be offered on a spoon).
> - Try food with a stronger smell.
> - Offer only very small pieces at a time.
> - Scattering tiny treats for the cat to find can sometimes make food more interesting.
> - Vary the treats.
>
> If the cat appears obsessed by the food, if he tries to grab it or shows greater interest in the food and much less interest in interacting with the trainer or in performing the desired behaviour:
>
> - Use a less appealing food e.g. dry kibble.
> - Place the treat in a pot, drop it onto the ground or give using tongs or on a long-handled spoon to prevent the cat from biting or clawing at hands.
> - Use another form of training reward.
>
> **Stroking/grooming**
>
> - Will not work as a reinforcer if the cat does not like to be touched, or will only occasionally accept or enjoy being petted.
> - Can work well for cats that enjoy human interaction, especially cats that enjoy being stroked and/or groomed.
> - The head and/or side of the face is usually where cats most like to be stroked (Ellis *et al.*, 2015).
> - Providing a brush or similar that the cat likes to rub against can work very well for some cats.
>
> **Play**
>
> - Playing a game with a wand-type toy can work well as a reward and reinforcer for a young cat or an older cat that enjoys play.
> - Playing is not advisable to use as a reward if the desired behaviour requires that the cat remains calm and relaxed.

directly necessary for our survival or well-being. But because money has become very strongly associated with the things that we do need or want it has become equally desirable.

For animals, irrelevant stimuli, especially sounds, can become associated with primary reinforcers such as food and become just as desirable and, if used correctly, can be a very effective training tool.

> **Box 7.2. Schedules of reinforcement.**
>
> **Continuous reinforcement**: the behaviour is rewarded every time it occurs.
> **Partial (intermittent) reinforcement**: the behaviour is rewarded intermittently.
>
> **Ratio schedules**
>
> **Fixed ratio**: the subject is only rewarded if he has responded correctly a set number of times.
> **Variable ratio**: the number of correct responses varies between rewards.
>
> **Interval schedules**
>
> There is a time interval between rewards.
> **Fixed interval**: the subject is only rewarded for a correct response that occurs a set time after the last correct response.
> **Variable interval**: the time between each reward for a correct response varies.
>
> **What is best for training?**
>
> Continuous reinforcement is advisable at the start of training to allow an initial strong association to be made between the behaviour and the reward. But one problem with continuous reinforcement is that when the rewards stop the behaviour will quickly extinguish, i.e. no longer be performed, whereas behaviours that have been trained using a partial schedule of reinforcement, particularly a variable ratio, have been shown to be much more resistant to extinction. In other words, they persist longer in the absence of reward.
>
> Therefore, once an initial strong association has been made, the trainer is best advised to change from rewarding the cat every time it does the right thing, to offering rewards intermittently but with no set pattern.

Basic principles of secondary reinforcement training

- The stimulus, e.g. a sound, is repeatedly paired with a primary reinforcer, e.g. a food treat: sound first and then the food treat.
- The animal not only learns to anticipate the food as soon as it hears the sound but, due to the strong association, the sound alone can act as a reward, but only if it continues to be associated with the food treat.
- As soon as the animal performs a correct response, the trainer activates the secondary reinforcer (e.g. the sound), which 'marks' the behaviour and then the primary reinforcer is given.

Advantages of this training:

- When using a primary reinforcer there can be a delay between the performance of the required behaviour and the delivery of the reward. This can result in the desired behaviour being missed or another behaviour being unintentionally rewarded and reinforced.

- When using a secondary or conditioned reinforcer this can be used 'instantly' at the same time as the animal performs the behaviour, or immediately after, thereby signalling clearly to the animal exactly what is wanted.
- This can also be convenient for the trainer because he or she only needs to have immediate access to the secondary reinforcer.

Disadvantages of this training:

- The trainer's timing needs to be good to avoid marking and unintentionally reinforcing the wrong behaviour.
- A 'clicker' is commonly used for secondary reinforcement training of dogs and other animals but this may not be so suitable for cats (see Box 7.3).

Luring

Obviously, an animal needs to perform a behaviour before we can reinforce it. Physically forcing the animal into a place or position does not teach it anything because it needs to physically carry out the action itself in order to learn how to do it again. Some behaviours occur spontaneously, but then we might have to wait around a long time for the cat to do what we want it to. The chances that the behaviour will be performed when we start training can be increased by encouraging (luring) the cat to do what we want by using treats or other enticements (Fig. 7.2).

Practical feline examples

- We may want to teach the cat to lie down and relax on a specific mat or cat bed. We may start by encouraging the cat to go in the right direction by throwing treats onto, or close to, the bed or by luring the cat onto the mat using wet food on the end of a long spoon.
- Play and toys can also be used as lures. For example, if training a cat to use a cat flap a favourite toy can be used to lure the cat through the cat flap from one side to the other. Play, however, would not be the best thing to use in the previous example if the intention is to teach the cat to relax.

Fig. 7.2. Luring can be adapted to be used as a hand signal that can also be later paired with a verbal 'command', e.g. 'sit'.

> **Box 7.3. What is the best conditioned reinforcer to use with cats?**
>
> The stimulus should be one that is easy and quick to impart and is sufficiently salient for the cat. Unless the cat is deaf, a sound is often the best choice because the cat does not need to be looking at the trainer at the time. Sounds can also be of short duration and easily 'turned on and off'.
>
> The sound also needs to be one that the cat is unlikely to hear often in everyday situations. A clicker, a device that produces a loud audible 'click' is often used for this type of training with other animals; however, there can be disadvantages in using a clicker with cats:
>
> - Cats can be comparatively sound sensitive and a clicker may be too loud and potentially frightening for some.
> - Using a clicker or any other sound-producing device may require some practice and coordination to get the timing correct.
>
> A verbal signal, such as a specific word or sound can be a better choice when training cats:
>
> - The word or sound should be produced clearly, concisely and loud enough for the cat to be aware of it, without being too loud that it might startle the cat.
> - Avoid using sounds that could upset the cat, e.g. shhh or hissing noises.
> - Aim to make it sound the same each time you use it. Even if you choose a special word there could be various ways in which it may be pronounced.
> - Avoid using the cat's name or often used words or phrases, e.g. 'good boy/girl', that are commonly directed towards the cat, although single words from such phases, e.g. just the word 'good', can be used because this will sound quite different.

Shaping behaviour (successive approximation)

Even if we use a lure it is rare that the cat will do exactly what we want straight away, especially if we are expecting a reasonably complex behaviour. So in many cases we need to 'shape' the behaviour.

Each time the animal offers the behaviour it will be slightly different and the aim of 'shaping' is to concentrate on rewarding the actions that are closest to what we want. So, to begin with we would reward a behaviour that is 'near enough' to what we want and then each time the cat does something that is a bit closer to what we want, concentrating on only rewarding those behaviours and ignoring previous behaviours that were not so close, so that eventually the cat is being rewarded for doing exactly what we want it to.

Practical feline example

If we want to teach the cat to lie down and relax on a mat or cat bed we may shape the behaviour as follows:

- Reward initially for just going near the bed.
- Then reward only for putting at least one foot on the bed.
- Then reward only for standing on the bed.

- Then reward only for sitting on the bed.
- Then reward only for sitting on the bed for more than 10 seconds (gradually increase the time).
- Then reward only for lying down on the bed.
- Then reward only for lying on the bed for more than 10 seconds (gradually increase).
- Then reward only for signs of relaxation when lying on the bed.

The cue

This is the signal that precedes the behaviour and tells the animal what it is that we want it to do, more commonly known as the 'command'. This can be a verbal cue, e.g. a specific word or phrase, or a visual signal, e.g. a hand-gesture.

Exactly when to bring in the cue can vary depending on the behaviour we are hoping to reinforce and how we are training. If the training involves setting the cat up so that the chances of it performing the correct behaviour are very high, then the cue can be introduced at the very beginning of training. But, if the cat has many different options as to what behaviour it could perform, a common mistake can be to bring in the cue too early, before the animal has any understanding of what it is expected to do. This increases the risk that the cat will make an association with the cue and a behaviour other than the one we want, or it may never make the correct association, because the cue and the behaviour are not yet sufficiently well defined. In this case it can be better to wait until the cat has a good idea of what is expected of it and then bring in the cue as it performs, or is just about to perform, the desired behaviour. Signals used in luring, e.g. holding the hand as though about to give a treat, causing the cat to sit and look up, can also be adapted as cues and later paired with verbal cues.

However, if we are shaping a behaviour it may not be necessary to wait until the cat is performing the completed sequence, the cue can be introduced as soon as the cat has a general idea of what is expected of it.

Practical feline examples

- When teaching a cat to come to call, you would start with the cat reasonably close to you and lure it towards you with a food treat or other sufficiently rewarding enticement. In this case the cue (the signal used to call the cat) can be introduced very early on because there should be a very high chance that the cat will come to you to get the reward.
- If teaching a cat to sit 'on command', repeatedly saying 'sit' before the cat has any idea what is required of it is unlikely to be successful. It is better to wait until the cat does actually sit and then bring in the cue as it sits or is just about to sit. One way to lure a cat into a sitting position, however, is to hold a food treat in the air just above its head, so that the cat sits as it looks up at your hand holding the treat. This method of luring can then be adapted to be used as a visual cue that can later be paired with a verbal signal (Fig. 7.2).
- If teaching a cat to go and relax on a mat, the cue can be introduced as soon as the cat is actively going towards, and at least treading on, the mat. Shaping can then continue as described above.

Modifying (changing) unwanted behaviour

Punishment

Within the terminology of learning theory, if a behaviour and its consequence results in a decreased probability of the behaviour being repeated, it can be said that the behaviour has been punished. A punisher is the event that follows or accompanies a behaviour that results in punishment (Lieberman, 1993). A punisher is usually aversive, i.e. it is something that the animal would prefer to avoid, but for training or behaviour modification purposes it does not need to be painful or frightening. In fact, causing pain or fear in an attempt to punish an unwanted behaviour may cause or increase the development of more serious behaviours such as anxiety and aggression (Box 7.4).

Positive punishment

A positive punisher is an aversive event following a behaviour that effectively decreases the chances of the behaviour being repeated.

Practical feline examples

- The cat jumps into a neighbour's garden and is chased by the neighbour's dog. The cat learns to avoid that garden. Jumping into the neighbour's garden has been positively punished.
- The owner has wrapped 'cling film' around a chair leg that the cat normally scratches. When the cat attempts to scratch the chair leg, the feeling is unpleasant for the cat and the behaviour of scratching that particular chair leg is reduced and therefore has been positively punished.

Negative punishment

A negative punisher is an event following a behaviour that results in the end of a pleasurable experience and thereby decreases the chances of the behaviour being repeated.

Practical feline example

- The cat is enjoying sitting on the owner's lap and being stroked. He stretches and digs his claws into the owner's leg. The owner puts the cat down on the ground and moves away. The cat learns that digging his claws into his owner's leg results in the end of being stroked and feeling comfortable, so the behaviour has been negatively punished.

Non-reward

If the cat has come to expect a rewarding (reinforcing) outcome to a behaviour and then the expected reward stops, this may be sufficiently aversive to act as a

> **Box 7.4. The problems with punishment.**
>
> **Unintended associations**
>
> One problem with the use of punishment, or intended punishment for an unwanted behaviour is the risk that the cat may not associate the aversive event with the 'bad behaviour', but may instead make a completely different association.
>
> **Practical feline example**
>
> A cat is caught urinating on the carpet. The owner makes a loud frightening noise in an attempt at punishment. The cat associates the frightening event with another household cat who is nearby, resulting in fear and defensive behaviour towards the other cat and a breakdown in the relationship between the cats.
>
> **Timing**
>
> To increase the possibility (but not guarantee) that the cat will make the desired association, the aversive event needs to occur at the same time or immediately after the behaviour. Any delay increases the risk that the cat will associate it with a different behaviour.
>
> **Practical feline example**
>
> A cat jumps up on a work surface and steals food. The owner walks into the room and the cat jumps down to greet the owner. The owner shouts angrily at the cat for stealing the food but it is the behaviour of approaching the owner that is punished instead.
>
> **Intensity**
>
> To act as effective punishment, i.e. to decrease the likelihood that a behaviour will not be repeated, a punisher needs to be sufficiently aversive. However, if it is too aversive the more likely outcome is that it will cause fear, anxiety and stress without influencing the behaviour, owing to increased fear and arousal inhibiting learning (Teigen, 1994). To find the correct level of intensity can be extremely difficult because what is considered aversive is a subjective experience that can vary tremendously between individuals.
>
> **Practical feline example**
>
> Three cats in the same household are scratching the sofa. The owner attempts to punish the behaviour by squirting the cats with water whenever they scratch the sofa. For one of the cats, being squirted with water is sufficiently aversive that it avoids scratching the sofa (although it still scratches other items of furniture). For another cat, the squirt of water is not aversive enough and it continues to scratch the sofa. For the third cat, being squirted with water is highly traumatic and the experience causes it to become stressed and anxious, which negatively influences its relationship with the owner and the other cats and increases its scratching behaviour.

punisher. Also, if the behaviour is repeated without the expected reward, the continued lack of reinforcement can result in a decrease in the frequency and intensity of the behaviour and eventually the behaviour not being performed at all. This is known as **Extinction.**

Practical feline example

- The cat may keep opening the cupboard door where the treats are kept. But then the owner moves the treats to another cupboard out of the cat's reach. The cat continues to open the cupboard, but now that there are no treats in the cupboard, the behaviour is no longer rewarded and so the behaviour declines and eventually stops.

Extinction burst

If a behaviour is no longer rewarded, an 'extinction burst' may be seen before the behaviour becomes fully extinct. When this occurs, the behaviour increases or gets worse for a short time before declining.

Practical feline example

- Before the behaviour of opening the cupboard declines and eventually stops, the cat may initially increase the number of times it opens the cupboard, or become more frantic in its behaviour of opening the cupboard and searching for the treats, until it learns that the behaviour will no longer be rewarded.

Counterconditioning

Classical conditioning (see earlier) can result in a cat associating an otherwise unrelated stimulus with an emotional response, e.g. fear. Counterconditioning involves pairing the original conditioned stimulus with a different unconditioned stimulus that elicits a very different emotional response, i.e. pleasure.

However, counterconditioning alone is rarely successful, especially if the animal has become sensitized to experience fear on presentation of the stimulus. In most cases it is necessary to also desensitize and habituate the animal to the stimulus.

Desensitization

Introducing a stimulus that the cat has become sensitized to at a level that does not cause the undesirable emotional response, e.g. fear, and very slowly and gradually increasing the strength of the stimulus, allowing the cat to become less fearful and eventually habituate to it.

Practical feline example

The sight of the cat carrier may cause a cat to feel frightened due to previous associations with trips to the vets, etc. The fear may be reduced by the following steps:

- Leave the cat carrier out in a place where the cat will see it every day and become accustomed (habituated) to the sight of it.
- Teach the cat to associate the sight of the cat carrier with something good, e.g. tasty food treats, by dropping treats gradually closer to the carrier.
- When the cat no longer fears the carrier, further training can commence to teach the cat to enter the carrier willingly and without fear.

Redirecting behaviours or offering alternative targets

Behaviours that are innate and necessary for the cat's general well-being can also become a problem if the target of the behaviour is unacceptable in a human household. Another problem is when unwanted behaviours are self-rewarding and it is difficult or impossible to remove or prevent the reward. When this is the case, it is often best to offer the cat an alternative and more acceptable target for the behaviour.

Practical feline examples

- Scratching is a normal feline behaviour beneficial for health and welfare. But scratching on furniture is unacceptable for many cat owners. The problem may be resolved by offering a suitable scratch post and/or scratch pad and training the cat to use this instead.
- Cats can sometimes engage in unacceptable predatory-type play towards other cats or people. This can be resolved by re-directing the behaviour towards a toy instead.

Factors Influencing Learning

Motivation – learning new behaviour

There are several factors that can influence an individual's ability to learn, a major one being motivation. If a cat is 'not in the mood' to perform a behaviour, or is not interested in the potential reward, then performance of the behaviour is less likely to occur. Even if the cat does initially perform the behaviour, if the reward is something that the cat is not particularly interested in at the time there is less chance that the behaviour will be reinforced and therefore repeated. One of the difficulties in training cats can be in finding the right time and the right reward so that the cat is sufficiently motivated to perform the behaviour in the first place, and sufficiently interested in the reward so that the behaviour is reinforced.

Practical feline examples

- Cats spend a lot of time sleeping. A tired cat is less likely to want to engage in training.
- Trying to teach a cat to relax is much less likely to be successful when the cat is in the mood to play.
- The reward on offer can depend on what the cat most wants at the time. The motivation for food treats can be stronger when the cat is hungry, but if the cat is more interested in affection and attention at the time, then being stroked will be a stronger motivation and better reward. Likewise play may be the best reward when the cat wants to be active.

Motivation – modifying unwanted behaviour

When trying to change a previously learned behaviour, the motivation to perform the new behaviour needs to be stronger than the motivation to perform the old behaviour.

Practical feline examples

- The scratch post and/or pad offered to a cat that is scratching the furniture needs to be made more attractive to the cat than the furniture it is already scratching.
- A toy or game used to distract, or re-direct unacceptable predatory type play directed towards people or other animals, needs to be more 'fun' than the person or other animal that the cat is 'playfully' attacking.

Conditioned motivation

Motivation may be increased by the presentation of a stimulus already associated with a primary reinforcer. Weingarten (1983) taught rats to associate the sound of a buzzer with food. Food was then offered freely to the rats, without the buzzer, until they had eaten their fill and were no longer hungry and appeared uninterested in the food offered. But when they heard the buzzer they would start to eat again, despite being full.

Practical feline example

- It can be a good idea to prepare a 'training toolbox' containing everything necessary for training including all different types of rewards (Bradshaw and Ellis, 2016). If previous training sessions have been fun and sufficiently rewarding for the cat, the sight of this box can then signal that training is about to commence and increase the cat's motivation for training.

Discriminative stimulus

A cat may learn that an event will only occur in the presence of a specific stimulus. This may be separate from, or in addition to, the stimulus with which we intend the cat to associate the event.

Practical feline example

- A squirt of water from a water pistol may be used to deter an invading cat from entering the garden, the aim being that the cat learns to associate coming into the garden with getting wet. But the cat may still come into the garden when the person with the water pistol is not around because it is their presence that the cat has learned to associate with being squirted with water.

Overshadowing

More than one stimulus may be present during classical conditioning. When this occurs the animal will make the association with the stronger, or more salient stimulus.

Practical feline example

- When teaching a cat to come to call an owner may crouch down, hold a treat out towards the cat and softly call the cat's name. The cat may go to the owner, but the visual cue of the owner crouching down and holding out the treat may be more salient, and therefore overshadow the sound of the cat's name being called. The result being that the cat does not come to call when he cannot see the owner, or if the owner does not crouch down and hold out a treat.

Superstitious behaviour

During training (operant conditioning) an animal may perform a response in addition to, or instead of, the behaviour we intend to reinforce. If reward follows the performance of this response the animal may believe that this is the behaviour that elicited the reward.

Practical feline example

- When training a cat to sit on command, the cat may also lift its paw after being given the command to 'Sit'. The cat may then learn that the command 'Sit' means 'lift paw' or 'sit and lift paw'.

Context-specific learning

When an animal learns a response, it will make associations not only with the direct consequence of the behaviour but also with the set of circumstances surrounding the event or place where learning took place. If learning takes place in a variety of different contexts, the behaviour and learning will become generalized to most other contexts. But if learning only occurs in one context, the same response will be much less likely to occur in other contexts.

Practical feline examples

- If a cat has had a frightening experience in the kitchen, but nowhere else in the house, it is more likely to exhibit fear and vigilance in the kitchen and less likely to in other areas of the house.
- If the cat has been trained to come to call while it is in the living room and nowhere else, it will be less likely to respond when called from other areas of the house or garden.

Learned helplessness

An animal may eventually 'give up' trying to avoid an aversive stimulus if previous attempts have been unsuccessful.

Practical feline example

- A cat in a restricted situation, e.g. a shelter, cattery or hospital cage, might be highly fearful of the surrounding environment and attempt to hide. If there is nowhere provided for the cat to hide and it is in this situation for a reasonable length of time it may simply give up trying, even if provision to hide is later offered.

Health and cognitive abilities

Learning can be significantly influenced by the cat's physical and mental health. Cognitive dysfunction syndrome (see Chapter 6) will reduce an affected cat's ability to learn, while pain, discomfort or just feeling unwell can make a cat less willing and able to engage in training. Intended reinforcements will also be less attractive and rewarding for a cat in poor health. Internal pain or discomfort can also become associated with unrelated stimuli or behaviour.

Practical feline examples

- A cat with cystitis may experience pain when attempting to urinate in the litter tray and learn to associate the litter tray with pain, leading to a breakdown in house training.
- A cat may be less willing to 'come to call' if it is in pain or discomfort and does not want to move, or if it has no appetite or interest in the food treat that it may get when it goes to the owner.

References

Bradshaw, J. and Ellis, S. (2016) *The Trainable Cat: How to Make Life Happier for You and Your Cat.* Basic Books, New York.

Davis, M. (1974) Sensitization of the rat startle response by noise. *Journal of Comparative and Physiological Psychology* 87, 571–581.

Ellis, S.L.H., Thompson, H., Guijarro, C. and Zulch, H.E. (2015) The influence of body region, handler familiarity and order of region handled on the domestic cat's response to being stroked. *Applied Animal Behaviour Science* 173, 60–67.

Lieberman, D.A. (1993) *Learning, Behavior and Cognition*, 2nd edn. Brooks/Cole, Belmont, California, USA.

Pavlov, I.P. (1927) *Conditioned Relexes*. Oxford University Press, London. Cited in: Lieberman, D.A. (1993) *Learning, Behavior and Cognition*, 2nd edn. Brooks/Cole, Belmont, California, USA.

Teigen, K.H. (1994) Yerkes-Dodson: A law for all seasons. *Theory & Psychology* 4, 525–547.

Weingarten, H.P. (1983) Conditioned cues elicit feeding in sated rats: a role for learning in meal initiation. *Science* 220, 431–433.

Part Two
Practical Feline Behaviour

Introduction

This section of the book provides specific advice on how to improve feline welfare and help to prevent unwanted behaviour issues.

Although the chapters are headed as 'Advice for...', readers should not feel that they are restricted to reading only the chapter that is targeted towards their own specific role as a cat carer, because the advice contained in the other chapters should also prove to be informative and provide an overall picture of how to aim for good feline behavioural welfare.

8 Advice for Breeders

Producing adorable kittens that can be sold for profit can seem like a very attractive prospect but in reality breeding is not always easy or profitable. Pregnancy and lactation can be a physically stressful time for a female cat and significant knowledge, time and financial input is required to ensure that a queen and her kittens remain healthy. Even then, complications can arise requiring additional veterinary care, possible loss of the kittens, and significant negative impact on the queen's physical health and welfare.

The reader is also referred to Chapter 5, Kitten to Cat, for information relevant to this chapter. Anyone considering breeding from their cat is best advised to research well beforehand by asking for advice from experienced and reputable breeders and from their veterinary surgeon or nurse/technician. Advice can also be obtained online from reputable organizations such as the Governing Council of the Cat Fancy (GCCF) (https://www.gccfcats.org/Breeding-Information) and International Cat Care (https://icatcare.org/advice/breeders).

The Responsibility of the Breeder in the Prevention of Behaviour Problems

Much of what shapes the behaviour of a cat can be linked to events and experiences it had as a very young kitten, and to influences on its development before birth. A large part of the responsibility for producing kittens that grow up to be physically and behaviourally healthy cats therefore lies with the breeder. This responsibility should be the same regardless of whether the kittens are pedigree or non-pedigree.

Selection of Queen and Stud

- Aspects of behaviour such as innate levels of confidence or timidity can be inherent (Turner *et al.*, 1986; McCune, 1995). Therefore, it is important that stud males and breeding queens are both of good temperament, confident, friendly and well socialized.
- The breeding queen should not have any undesirable behavioural issues. Kittens learn by observation of their mother's behaviour and are likely to exhibit the same behaviours. For example, kittens born to mothers that have house-soiling issues and do not, or rarely, use a litter tray are less likely to learn how to use a litter tray themselves, and may continue to have house-training issues even as adults.

- Both tom and queen should be in good health, free from known hereditary faults and fully vaccinated:
  - Infectious diseases and congenital disorders can be passed onto the kittens.
  - Pregnancy and lactation are physically demanding and will further weaken a cat that is not in the best of health.
  - Ill health can increase pre-natal stress (stress experienced by the queen during pregnancy), which can have a direct and potentially damaging influence on the behavioural development of the kittens (Weinstock, 2008).
- Both the queen and tom should be physically and behaviourally mature – at least 18 months to 2 years of age:
  - An immature queen will be less able to cope with the physical and emotional stresses associated with pregnancy and lactation and thereby be more likely to experience pre-natal stress.
  - A young mother is more likely to demonstrate poor mothering skills.
  - There is an increased risk that an immature queen will abandon or even attack the kittens.

Keeping a stud cat

Some stud cats are kept as family pets within the home, but many are confined to outdoor pens to avoid damage to the home from urine spraying or unintentional matings with entire queens that may also be kept in the house. But isolation and confinement can seriously limit the stud's opportunity to engage in behaviours such as exploration, predatory play and social interactions with people, all of which can be essential for his well-being and both behavioural and physical health. If there is no other option than to keep the stud cat in a separate pen, the following suggestions may help to improve his welfare and reduce negative impacts on his behaviour:

- Allow occasional but regular access to a larger and more interesting area of the garden by using cat-proof fencing.
- Spend some time each day with him and, as long as he is not aggressive, allow other people to interact with him (if he is aggressive consideration should be made as to whether he is a suitable cat from which to breed). Even if he was well socialized with people as a kitten, fear and mistrust of people can develop and increase if he is kept isolated and away from human interactions.
- Keep him well away from other entire males. The sight, sound and smell of other entire tomcats is likely to increase his sense of competition and overall stress.
- Provide plenty of environmental enrichment (see Appendix 1) within his pen, such as food foraging/puzzle toys and furniture and/or shelving that provide access to surfaces at varying levels, thereby giving him increased opportunity to climb and explore.

Pregnancy

Minimizing pre-natal stress

Stress experienced by a queen when pregnant will cause the production of stress hormones that can cross the placenta and have potentially damaging effects on the behavioural and physical development of the unborn kittens (Weinstock, 2008). It is therefore very important to minimize stressors for a pregnant cat.

- Prior to conception and all through pregnancy ensure that the queen is physically, emotionally and nutritionally healthy. Regular health checks by your veterinary surgeon or nurse/technician before she is mated and during pregnancy are advisable (Fig. 8.1).
- Mothers that are poorly nourished can produce kittens with developmental abnormalities (Simonson, 1979, cited in Bateson, 2000; Gallo *et al.*, 1980). Poor health and insufficient nourishment when nursing can also result in the mother weaning the kittens earlier and increased aggression by the mother towards the kittens during weaning. Sufficient high-quality nutrition throughout pregnancy and lactation is therefore very important:
  - Ensure that she has easy access to both food and water. Provide additional bowls and position water dishes away from food.
  - Feed little and often – her appetite is likely to increase but she may not be able or willing to eat too much at one time.
  - Feed a diet that meets her increased nutritional requirements, such as a well-balanced prepared kitten food.
  - Research has shown that kittens will have a preference for the same food eaten by their mother during pregnancy and lactation (Becques *et al.*, 2009; Hepper *et al.*, 2012), therefore it is advisable to feed her the same food as will be offered to the kittens during weaning.
  - If you are unsure as to the correct diet to feed, ask your vet or nurse/technician for advice.
- Avoid conflict situations with other pets:
  - Reconsider breeding from a queen that is already fighting with another cat or may feel threatened by another household pet. Do not breed from her until the situation is resolved.
  - Do not introduce new pets to the household, especially other cats, or dogs, during the pregnancy or while she is nursing her kittens.
  - Avoid or reduce any overcrowding or competition for resources (see Appendix 3).

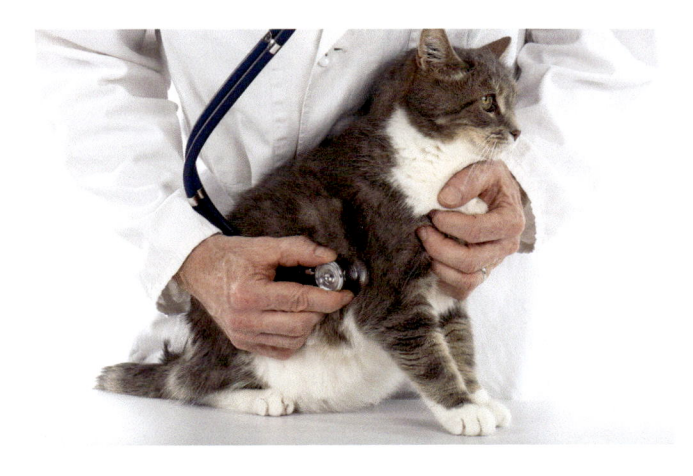

Fig. 8.1. Poor health can be a cause of pre-natal stress. Regular health checks before conception and during pregnancy are advisable.

- Avoid major environmental changes such as:
  - Re-decorating.
  - Building work.
  - Moving to a new house.
  - Time spent in a cattery or environment away from home.
- Avoid changes to the social environment such as:
  - People moving in or out of the home.
  - Frequent and/or long-term visitors.
  - New pets, or visiting animals.
  - Holidays or any situation that involves owners or usual carers spending considerable time away from home.
  - Changes to normal husbandry routine.
- Provide easy access to litter trays or other elimination areas. Extra litter trays may be required as she becomes less mobile.
- Provide easy access to comfortable resting areas and safe hiding places. As pregnancy increases and her mobility decreases she may not be able to reach high places where she may have previously felt safe.
- Avoid subjecting her to sudden changes in ambient temperature:
  - Keep her living quarters comfortably warm, but avoid over-heating.
  - Do not shut her outside in cold weather or extreme heat without access to cool shade.
- Allow her to have control over her environment and social interactions:
  - Do not restrict her access to places, people or other animals with which she wishes to interact.
  - Allow her to initiate interactions. Do not force interactions on her, and avoid picking her up or excessive stroking or cuddling, especially by strangers or people to whom she is not closely attached.

Preparation of the nest site

As her time to give birth approaches, the queen will try to find what she considers to be the best place to give birth (Fig. 8.2).

- Provide her with two or three different nest site options:
  - Give her a choice of locations and the option of being close to or at a distance from family members.
  - A few days or weeks following the birth it is normal for the mother to move the kittens. If she is provided with other suitable nest sites there should be an increased chance that she will move the kittens to one of these. Also, the more suitable the nest site, the less frequently she is likely to move the kittens.
- Nest sites should be in a quiet area where there is minimal risk of disturbance.
- The area should be dry, warm and free from drafts.
- Important resources, i.e. food, water, litter trays, etc. should be nearby and easily accessed but separate from each other.
- Bright lighting should be avoided; the area should preferably be in permanent semi-darkness.

Fig. 8.2. Ensure good provision of nest sites (at least two) that are warm, quiet, away from bright light and semi-enclosed without restricting the queen's movements.

- The nest site should contain clean, soft comfortable bedding that is also safe for the kittens.
    - Avoid loose knit or material containing holes that the kittens could get entangled in.
    - Avoid material that could easily disintegrate or release loose fibres.
    - Avoid any material that could potentially contain chemical toxins, for example if it has previously been used for other purposes such as cleaning or been stored close to harmful chemicals, such as in a garage or garden shed (ISFM, 2017).

Parturition

Signs that giving birth is imminent

- Decreased appetite.
- Increased grooming of the ventrum and anogenital area.
- Panting.
- Increased vocalizing, including purring.

During parturition

- Keep a close watch to ensure that there are no problems; if concerned contact your veterinary surgery, otherwise keep disturbances to a minimum.
- Keep the area in semi-darkness, warm and quiet.

Pre-weaning Period

- The mother will be reluctant to leave the kittens at this stage so ensure that her important resources, i.e. food, water and litter trays, are close by and can be easily accessed.

- During the first week or so after birth, ensure that mother and kittens are disturbed as little as possible. Severe or frequent disturbances can severely disrupt the relationship between mother and kittens and might result in the mother abandoning or even attacking the kittens. Aggression towards people and other animals might also be more likely if she feels that her kittens are under threat.
- Do not confine the mother to the nest site or severely restrict her movements. It is normal for the mother to move the kittens to a new nest site at least once during the pre-weaning period. Preventing her from doing so can cause unnecessary and damaging stress.

Weaning

- Natural weaning will normally commence between 3 and 4 weeks of age.
- Kittens will have a 'pre-programmed' preference for food that has the same flavour as that eaten by their mother during her pregnancy and lactation (Becques *et al.*, 2009; Hepper *et al.*, 2012). Therefore, weaning can be made easier by offering the kittens a softened version of the same solid food that their mother has been eating. If the mother, during pregnancy and lactation, and then later the kittens are offered a wide variety of different foods and flavours, this can help to reduce the likelihood of the kittens becoming 'fussy eaters' as adults.
- As weaning progresses, the mother cat will naturally try to prevent the kittens from nursing by adopting postures that deny them access to her teats and by generally keeping out of their reach. To aid this process, provide her with resting areas from where she can still watch over her kittens without them being able to get to her.
- If the queen does not have outdoor access to hunt and bring prey back for the kittens, provide small, safe, soft toys that the kittens can play with in the place of real prey.
- Do not interfere with the natural weaning process, unless unavoidable. A mother cat may appear to be aggressive towards her kittens when they attempt to feed from her at this time, but unless she is actually causing them harm, this is normal. A queen is, however, more likely to be aggressive towards her kittens if she is underweight, poorly nourished or if there are a large number of kittens in the litter. Ensuring that she is healthy and well nourished can therefore help to prevent aggression towards the kittens.

Hand-rearing

- Do not separate the kittens from their mother unless it is absolutely unavoidable. Doing so can be a severe stressor for the kittens that can have damaging life-long effects on their health and behaviour (Fig. 8.3).
- Seek advice on hand-rearing from a reputable and knowledgeable source (e.g. https://icatcare.org/advice/hand-rearing-kittens).

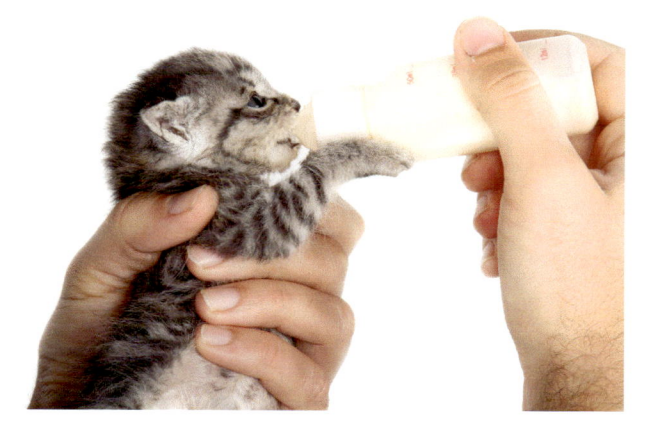

Fig. 8.3. Do not separate the kittens from the mother unless unavoidable. Frustration is a normal part of the natural weaning process that hand-reared kittens are less likely to experience, and lack of this experience can result in behavioural problems later in life.

Natural weaning vs weaning following hand-rearing

Weaning by the mother cat is a process that naturally causes a sense of conflict and frustration for the kittens. This is because their availability to suckle from the queen decreases and her nursing behaviour becomes unpredictable. She will intermittently and increasingly deny them access to feed from her by keeping her distance and by adopting resting postures incompatible to nursing; she will even get up and walk away whilst they are feeding. The frustration experienced by the kittens at this time is an important part of the weaning process in two respects:

1. It encourages the kittens to seek food elsewhere, and increases their desire to attempt and learn predatory behaviour.
2. It teaches the kittens how to cope with the emotion of frustration when it is experienced in later life. Cats that are unable to cope with this emotion are more likely to experience stress and can have an increased propensity to become aggressive when expectations are not met or anticipated rewards are delayed.

Frustration will also be experienced by the kittens during their early attempts to catch and kill live prey presented to them by their mother.

In contrast, kittens that have been hand-reared and weaned by a human caregiver are less likely to experience the same level of frustration as mother-reared kittens and may be more likely to exhibit frustration-related behavioural problems in later life. The following advice for carers of hand-reared kittens may help to prevent this:

- Do not offer solid food by hand as this can teach the kittens to associate human hands with food. It is better to offer solid food mixed in with formula milk in a shallow dish and gradually reduce the amount of milk and increase the solid food.
- As weaning commences, it is advisable to become more unpredictable in the feeding schedule and do not feed on demand.
- As the kittens start to take solid food, or food mixed with formula, start to withdraw the feeding bottle when there is still a small amount of formula left in the bottle.
- Once the kittens are eating solid food on a more regular basis, start to introduce simple puzzle feeders and food foraging games (see Appendix 1) (Bowen and Heath, 2005; ISFM, 2017).

Early Experience

Because the feline sensitive period for learning is between approximately 2 and 7 weeks of age (Karsh and Turner, 1988; Knudsen, 2004), (see Chapter 5) the responsibility to ensure that kittens are well prepared for their future life sits squarely with the breeder or carer of young kittens.

Habituation

A fear or startle reaction is normal in response to novel or unidentified stimuli, but if an animal were to react fearfully to every sight or sound it experienced the result would be increased stress and reduced welfare. Habituation is a natural process whereby an animal learns that most everyday sights, sounds and smells are harmless and therefore there is no need to respond to them. Habituation to everyday stimuli occurs more readily and easily during the sensitive period.

- Kittens should be raised in an environment that is similar to where they are intended to spend the rest of their lives. Kittens intended to be household pets should be raised in a family home where they will regularly experience and become accustomed to everyday household sights, sounds and smells.
- Sounds that kittens may encounter as adults, but are not a part of their everyday surroundings, e.g. thunderstorms and fireworks, can also be introduced in the form of recorded sound effects. These sounds can often be loud and frightening in real life and if recordings are played at a realistic level, the result can be to cause increased fear and reaction (see 'Sensitization' in Chapter 7) rather than habituation. Therefore, these should initially be played at a very low level and the volume very gradually increased as the kittens become accustomed to them.

Socialization

This is the process whereby a young animal develops social attachments and learns to recognize and to develop appropriate social behaviour with members of its own species and with members of other species with which it lives. Kittens that do not experience sufficient or appropriate socialization during the sensitive period are unlikely to be suitable as domestic pets.

Socialization with other cats

Frequent and regular positive encounters with adult cats, in addition to the mother cat, can increase tolerance of other cats as an adult (Kessler and Turner, 1999). Raising the kittens in an environment with other friendly and well-socialized adult cats can therefore be advantageous. It is important, however, to ensure that these cats are not only friendly towards the kittens but also that they have a good relationship with the mother cat.

Socialization with other animals

Regular positive encounters with other species that the kittens may later go on to live with, such as dogs, are also advisable. As with other cats, a dog in the household where the kittens are raised should be well socialized and friendly with cats and people, be non-threatening to the kittens and well accepted by the mother cat. However, it is also important to be aware that, owing to the tremendous variation between different dog breeds and types, the full effect of early socialization with dogs might be limited to dogs of a similar size and breed type.

Socialization with people

Positive early socialization with a variety of people is essential if the kittens are intended to be companion animals. It is also important that this is commenced prior to 7 weeks of age and is done correctly. Too much handling, rough or inappropriate handling or any interaction with the kittens that causes the mother cat to feel that they may be at threat is more likely to instil fear and avoidance. The following, adapted from Casey and Bradshaw (2008), can be used as a guide as to how the kittens should be handled.

Under 2 weeks of age:

- To avoid distressing the mother cat, the kittens should only be handled minimally by the person the queen is most comfortable with.

From 2 weeks of age:

- Handling should be restricted to people that the mother cat knows and is happy with.
- Talk to the kittens whilst stroking them.
- Pick up each kitten individually and cradle in the hands for a few seconds before replacing back with mother and littermates.

From 3 weeks of age:

- Still restrict handling to people that the mother cat already knows and trusts.
- Talk to and stroke the kittens.
- Pick each one up and hold for approximately 30–60 seconds, stroking all over the body. But if the kitten or mother cat appears distressed put the kitten back down or allow it to sit on your lap rather than being held in your hands.

From 4 weeks of age:

- Allow other people to handle the kittens; however, strangers should interact with the mother cat first. If she appears fearful or aggressive towards them, allow time for her to become accustomed to them before allowing them to handle her kittens. If she remains fearful and defensive, handling of the kittens should occur away from the mother but within the presence of other littermates.
- Gradually increase length of handling to 2–3 minutes at a time.

From 5 weeks:

- Increase handling to 5 minutes or more and introduce playing with a wand toy or similar.
- Repeat several times daily, with a variety of people.

From 6 weeks:

- As at 5 weeks, but whilst being handled, each kitten is occasionally carried a short distance away from the queen and littermates.
- Playtime with a toy is increased.

At 7–9 weeks:

- As at 6 weeks but play and handling sessions away from the mother and littermates are increased in length and frequency.

Frequency of handling and other considerations

- Kittens need a lot of sleep, they also need time to play with littermates and explore their environment. It is therefore important to allow them to engage in these activities and do not disturb them if they are sleeping.
- After 2–3 weeks of age the kittens should be handled daily for a minimum of 30–60 minutes in total, divided into short periods.
- They should be handled regularly by a minimum of four different people, preferably more.
- They should be handled by people of varying ages, gender and appearance (Fig. 8.4).
- Handling by strangers, especially children, must always be supervised, and advise visitors how to correctly handle and interact with the kittens.
- Always consider the kitten's health and possible risks of infection:
    - Encourage people to wash their hands before handling the kittens.
    - A clean blanket or towel draped over the person's lap can help to minimize contact with a stranger's clothes, plus it can act as an additional 'comforter' for the kitten if it contains familiar scents.
    - Any animals, especially other cats, that come into contact with the kittens must be healthy and fully vaccinated.

Fig. 8.4. Kittens between the age of 2 and 7 weeks of age should be gently handled by a variety of people. Handling, especially by young children, must be supervised.

Maternal Aggression

Occasionally a queen may become uncharacteristically aggressive towards people and/or other animals that approach her kittens. This can sometimes be lessened by allowing her to become accustomed to the person or other animal before they approach her kittens, but if she remains fearful or aggressive then interactions with the kittens should occur away from the mother but in the presence of littermates. This is because the kittens will learn by observation of their mother's reactions so are more likely to also become fearful and potentially aggressive. It is important, however, that they stay with littermates because their presence can help to increase confidence.

A queen that has shown maternal aggression should not be bred from again because she is very likely to repeat the behaviour. There is also a high possibility that her female offspring may demonstrate the same behaviour if they have kittens (Meaney, 2001) (Fig. 8.5).

Education of New Owners

The following two chapters contain information that can be passed on to new owners to further help prevent the development of behaviour problems. The importance of healthcare and vaccinations should also be discussed with owners and it should be emphasised that any pet cat not specifically intended for breeding must be neutered (see Appendix 4). Unfortunately, many non-pedigree pet kittens are the result of unintentional matings between unneutered pets and feral or unowned toms. When this occurs, there is usually no way of ensuring the physical health or temperament of the father. Also, the majority of 'unplanned pregnancies' occur during the queen's first season when she is still very young and often not much more than a kitten herself.

Fig. 8.5. A queen that demonstrates severe maternal aggression should not be bred from again. Also, her kittens should be neutered because they are likely to exhibit the same behaviour.

In these cases, the harmful effects of pregnancy and lactation on the mother's health, and on her kittens' behavioural health and welfare, can be considerable.

References

Becques, A., Larose, C., Gouat, P. and Serra, J. (2009) Effects of pre- and post-natal olfactogustatory experience at birth and dietary selection at weaning in kittens. *Chemical Senses* 35, 41–45. DOI: 10.1093/chemse/bjp080.

Bowen, J. and Heath, S. (2005) *Behaviour Problems in Small Animals: Practical Advice for the Veterinary Team.* Elsevier Saunders, Elsevier, Philadelphia, Pennsylvania, USA.

Casey, R.C. and Bradshaw, J.W.S. (2008) The effects of additional socialisation for kittens in a rescue centre on their behaviour and suitability as a pet. *Applied Animal Behaviour Science* 114, 196–205.

Gallo, P.V., Wefboff, J. and Knox, K. (1980) Protein restriction during gestation and lactation: development of attachment behaviour in cats. *Behavioral and Neural Biology* 29, 216–223.

Hepper, P.G., Wells, D.L., Millsopp, S., Kraehenbuehl, K., Lyn, S.A. and Mauroux, O. (2012) Prenatal and early sucking influences on dietary preferences in newborn, weaning and young adult cats. *Chemical Senses* 37, 755–766.

ISFM (2017) *How to Reduce Prenatal Stress. Module 3: Reproduction, Behavioural Development and Behavioural Health in Kittens.* International Society of Feline Medicine Advanced Certificate in Feline Behaviour. Available at: https://icatcare.org/learn/distance-education/behaviour/advanced (accessed 25 January 2018).

Karsh, E.B. and Turner, D.C. (1988) The human-cat relationship. In: Turner, D.C. and Bateson, P. (eds) *The Domestic Cat: The Biology of its Behaviour.* 1st edn. Cambridge University Press, Cambridge, UK.

Kessler, M.R. and Turner, D.C. (1999) Socialization and stress in cats (*Felis silvestris catus*) housed singly and in groups in animal shelters. *Animal Welfare* 8, 15–26.

Knudsen, E.I. (2004) Sensitive periods in the development of the brain and behaviour. *Journal of Cognitive Neuroscience* 16, 1412–1425.

McCune, S. (1995) The impact of paternity and early socialization on the development of cats' behaviour to people and novel objects. *Applied Animal Science* 45, 109–124.

Meaney, M.J. (2001) Maternal care, gene expression, and the transmission of individual differences in stress reactivity across generations. *Annual Review of Neuroscience* 24, 1161–1192.

Simonson, M. (1979) Effects of maternal malnourishment, development and behaviour in successive generations in the rat and cat. *Malnutrition, Environment and Behavior*. Cornell University Press, Ithaca, New York, pp. 133–160. Cited in: Bateson, P. (2000) Behavioural development in the cat. In: Turner, D.C. and Bateson, P. (eds) *The Domestic Cat: The Biology of its Behaviour*, 2nd edn. Cambridge University Press, Cambridge UK, pp. 9–22.

Turner, D.C., Feaver, J., Mendl, M. and Bateson, P. (1986) Variations in domestic cat behaviour towards humans: a paternal effect. *Animal Behaviour* 34, 1890–1892.

Weinstock, M. (2008) The long-term behavioural consequences of prenatal stress. *Neuroscience and Biobehavioral Reviews* 32, 1073–1086.

9 Advice for Prospective Cat Owners

Pet ownership can be beneficial for both the human carer and the pet animal. For people, owning a pet can provide companionship, enjoyment, entertainment and even health benefits (see Box 9.1), while the companion animal can benefit from being provided with a reliable food source, shelter, security, veterinary care and protection from predators and disease (Bernstein, 2007).

However, it can be easy for a pet–owner relationship to break down or to be a source of stress for either or both parties from the start. The blame is often put on the animal when behaviour problems or other difficulties arise, but the owner also has a responsibility to do the best that he or she can to avoid potential problems. The first step in this process is to be fully aware of the responsibilities involved in pet ownership and to make the correct choice of pet in the first place.

Is a Cat the Best Pet for You?

A cat can be the perfect pet for many people. Cats provide affection, companionship and often require less time and financial input than a dog. It is very wrong, however, to choose a cat simply because it appears to be an easy, or cheap, option. Owning a pet cat still requires a high level of commitment, and a potential owner must be fully aware of all the responsibilities that cat ownership entails (see Box 9.2).

A cat may not be your best choice of pet if any of the following are the main reasons you want a cat:

As a 'second choice' of pet or as a replacement for a dog

Dogs and cats are very different animals. Getting a cat on impulse when you really want a dog but are unable to have one is not a recipe for a good pet–owner relationship. It can be better to enjoy the company of dogs by volunteering at a local shelter or by fostering if you are unable to offer a long-term home for a canine companion.

Many dog lovers do also enjoy the company of cats and are equally committed to them, but if you have little or no experience of cats it is always best to find out as much as you can about them, and maybe even spend some time with cats at a local shelter or with friends' pets before acquiring your own pet cat.

As a small pet that requires little space

Some large dogs may actually require less space within the home than a small cat, especially if the cat is kept indoors with limited or no outdoor access. Cats are agile

> **Box 9.1. Health benefits of pet ownership.**
>
> - Increased relaxation.
> - Reduced anxiety.
> - Reduced stress.
> - Lowered blood pressure and heart rate.
> - Increased survival and increased longevity following a heart attack.
>
> (Friedmann and Thomas 1995; Bernstein 2007; Dinis and Martins, 2016; Kanat-Maymon et al., 2016.)

creatures who need sufficient space to run and climb. They also require areas to rest and places to hide if feeling threatened. So, as well as plenty of floor space, they also require access to vertical space, e.g. the tops of furniture, shelves, indoor cat trees, etc. (see Appendix 1).

Space is also required for food, water and litter trays, all of which need to be positioned well apart from each other. If there is more than one cat in the household, multiple separate feeding and water areas may be required, and because some cats prefer one area for urination and a separate area for defecation, even a single cat may require at least two litter trays positioned in separate locations. If there is more than one cat in the house the recommended number of litter trays to provide is one per cat plus one extra.

As a pet that requires little financial input

A non-pedigree cat or kitten might be acquired at little or sometimes no cost, but the care of a cat, regardless of its breeding can be expensive.

A potential cat owner should consider whether they are able not only to cover the cost of food, routine veterinary care, and other everyday costs such as cat litter, cat carrier, toys and so on, but also if they can cover the cost of cattery fees and unexpected veterinary fees for illness or injury. If not able to cover large unexpected bills, can you afford to pay for annual pet health insurance?

As a gift

No animal should be given as gift unless the recipient is able to make their own choice and is fully aware of and prepared for the responsibility that caring for the animal entails. Even children given the gift of a pet animal by a parent or guardian should be made aware of how much they are expected to contribute to the pet's care and welfare, although in the UK legal responsibility for the care of a pet animal rests with the parents or guardians until the child is 16 years old. This age of legal responsibility varies in different countries according to local jurisdictions.

As company for an existing cat

Despite being descended from an animal that lives a predominantly solitary existence, the domestic cat has developed the ability to be social with members of its own kind. They can certainly form close social bonds with other cats and cats in such a relationship do appear to benefit from it. However, there are many factors that can influence a cat's acceptance of another cat within its home (see Appendix 5) and sharing home and resources with another cat can be a source of stress rather than comfort for many. Correct introductions (also see Appendix 5) can increase the likelihood that a new cat may be accepted, but even then it can never be guaranteed that the cats will develop a relationship that is beneficial to both of them.

The same is true of cats that have lost a close companion. In a survey conducted in 1995 (cited in Caney and Halls, 2016), 217 owners of cats that outlived a feline companion reported on how the remaining cat reacted to the loss. The majority reported that the existing cat appeared unaffected, whereas some reported a positive reaction and others reported a negative reaction, describing behaviours that could indicate that the remaining cat was grieving.

Providing a new companion for a grieving cat may therefore seem like the best thing to do but, even if a cat has previously lived with and had a close relationship with another cat, he or she may not be so accepting of a 'stranger' and the introduction of another cat that could be regarded as a potential rival, or a lively and boisterous kitten that may be seen as a threat, can actually add to, rather than reduce, the stress associated with the loss of a close companion. See Appendix 6 for advice on how to help a cat that may be grieving the loss of a feline or human companion.

Making the Right Choice

Once you are certain that a cat is the right pet for you and you are aware of all that cat ownership entails, the next step is to make the right decision in finding the right cat for you.

Adult cat or kitten?

Rescue centres are full of cats looking for good homes and an adult cat can be no less of a rewarding and affectionate pet than a cat acquired as a kitten. Adult cats can be especially suitable for elderly people or anyone who may be less able or willing to cope with a lively and boisterous kitten. Taking on a pet also means taking on the responsibility for that animal's care for the rest of its life. If taking on a kitten, this can mean a commitment of 14–15 years or more (it is not uncommon for a cat to live into its early 20s). Adopting an older cat reduces the likely length of this commitment and so can be preferable for anyone who might not be able to commit to such a long duration of dedicated pet care (Fig. 9.1).

> **Box 9.2. The responsibilities of cat ownership.**
>
> - A pet owner is responsible for the safety, health, and welfare of any animal in their care. This responsibility should continue for the whole of the animal's life. If the pet is a cat this may be for 15 years or more.
> - The owner has responsibility to ensure that the cat is always provided with fresh clean water and a well-balanced diet that meets its individual nutritional requirements.
> - The owner is responsible for providing a safe and clean environment that provides protection from hazards, shelter from cold and wet, places to hide and comfortable, draught-free places to rest undisturbed. Sufficient space, comfort and diversity of surroundings should be provided to allow the cat to engage in normal behaviours. The cat must also have easy access to suitable toileting areas. If indoor litter trays are provided these must be cleaned and emptied regularly to reduce the risk of disease and associated behaviour issues.
> - The owner is responsible for providing that the cat has access to preventative health care as advised by their veterinary surgeon and appropriate professional veterinary treatment for disease or injury.
> - A pet cat that is not intended for breeding should be neutered before or very soon after reaching sexual maturity, to reduce the risk of unwanted pregnancies and to avoid health and hormonally influenced unwanted or antisocial behaviour issues, such as fighting, urine marking and excessive vocalization.
> - The owner is responsible for ensuring that the cat does not cause a nuisance or damage to other people, their pets or property.
> - The owner is responsible for ensuring that suitable care for the cat continues whenever they are unable to care for it themselves, for example due to sickness or holidays. They should also ensure that the person who will be looking after the cat fully understands its needs and any special requirements. The cat must never be left with anyone who might hurt or frighten it.
> - If an owner is unsure of any aspect related to the care and welfare requirements of their pet they must contact a suitably qualified cat care specialist, for example:
>   - Veterinary surgeon.
>   - Veterinary nurse/ technician.
>   - Animal behaviourist.
>   - Animal welfare organization.
>
> In the UK, failure to meet the welfare needs of a pet cat or causing unnecessary suffering might lead to prosecution under the Animal Welfare Act 2006.
>
> The Code of Practice for the Welfare of Cats 2006
> (https://www.gov.uk/government/uploads/system/uploads/attachment_data/file/69392/pb13332-cop-cats-091204.pdf)
>
> Cats and the Law: a plain English guide. The Cat Group
> (www.thecatgroup.org.uk/pdfs/Cats-law-web.pdf)

A major drawback of adopting an adult cat is that it may not be so easy to find out information regarding its previous and early life, and some cats relinquished to rescue centres are there because of behaviour problems that they exhibited in their previous homes. However, feline behaviour issues are very often linked to their environment, so behaviours that occurred in a previous home might not occur in a different environment.

Fig. 9.1. Adopting an older cat can be a better option for anyone who might be less able to cope with a lively kitten or might not be able to commit to continued care for the long lifespan of a cat.

Pedigree or non-pedigree?

If deciding on a pedigree cat or kitten, it is advisable to read up on the breed beforehand. It is important to know of any breed-specific requirements, for example long-haired cats will need regular grooming, and to be aware of any inherent disease conditions (for information see: https://icatcare.org/advice/cat-breeds/inherited-disorders-cats).

Male or female?

If cats are neutered prior to or very soon after reaching sexual maturity the differences between the sexes are minor, if any exist at all. Anecdotally, neutered males are reported to be more affectionate than females but there appears to be no scientific evidence published to support this.

There are, however, significant differences in the behaviour between neutered and entire (unneutered) adult cats and between entire males and females (Finkler *et al.*, 2011). Most of these behaviours are a part of reproductive behaviour (see Chapters 3 and 5) and are behaviours that most pet owners consider unwanted or unacceptable, such as urine marking, fighting and excessive vocalizing. Neutering is therefore essential, not only to prevent unwanted pregnancies and to improve the health and welfare of the individual cat, but also to prevent the development of these undesirable behaviours (see Appendix 4).

The Importance of Early-life Influences

The reader is referred to Chapters 5 and 8 for more detailed information.

What happens to a kitten before it is weaned and even before it is born can have lifelong influences on its behaviour and can make the difference between a kitten that grows up to be a friendly, relaxed and affectionate household pet or a difficult, nervous cat with a far greater prospect of exhibiting behaviour problems.

Personality traits such as levels of confidence or timidity can be inherited from the parents, and the health and welfare of the mother during pregnancy can have damaging effects on the behavioural development of her kittens. What a kitten experiences during the first couple of months of life is also of tremendous importance.

Socialization

Young animals, including kittens need to learn at a very young age which other animals are safe to be around. Kittens that have very little or no positive experience of people prior to 7 weeks of age are likely to consider humans as potentially threatening and remain fearful of people throughout their lives (Karsh and Turner, 1988; Knudsen, 2004).

Habituation

A fear reaction to novel or unidentified sights or sounds is a normal defensive behaviour that is essential for survival. A young animal needs to learn what sights and sounds are going to be a part of its everyday life so that it is not in a continual state of fear. Becoming accustomed to sights and sounds is a learning process known as habituation that also needs to occur when the kitten is very young and still with its mother.

What to Look For and What to Avoid

If acquiring a kitten

- Try to meet the kittens' mother, and father if possible (although this is rarely feasible because a pedigree stud cat may live some distance away and the father of a non-pedigree litter is usually unknown). The parents of the kitten should appear healthy, confident and friendly.
- Make sure that you see where the kittens have been raised. Kittens intended to be pets should be raised within the hub of a normal family home where they are likely to experience everyday household sights, sounds and smells. Kittens that have been raised in isolation, e.g. in an outhouse, external cattery, or separate room away from the rest of the household, might not have had the early experience to become sufficiently habituated to normal, everyday household sights and sounds and may be less suitable as household pets.

- You should be allowed to interact and handle the kittens. The breeder or carer may ask you to clean your hands first and/or may instruct you as to how to handle the kittens. But be wary if you are only allowed to view the kittens without interacting with them.
- It can be an additional bonus if there are other friendly adult cats around with which the kittens are able to positively interact. Kittens that have had positive experience with other cats when young are more likely to be tolerant of other cats when they grow up (Kessler and Turner, 1999).
- If you have a dog and/or children (or if there is a strong likelihood that both or either may become a part of your family), it is advisable that the kitten you choose has also had regular positive experience with dogs and/or children. It is also advisable to witness the interactions between the kittens and the dogs or children, to make sure that the kittens are not fearful in their company, and that the kittens are not being handled excessively or roughly. However, it is also important to be aware that, because of the tremendous variation between different dog breeds and types, the kittens are likely to only be accustomed to dogs of a similar size and breed type as those encountered at an early age (Fig. 9.2).
- It can be a good idea to visit your potential new kitten more than once. This is because there can be a lot of variation in a kitten's behaviour during the first 2–3 months of its life. At around 6–8 weeks of age fear and escape responses start to develop and a kitten may appear more timid at this time than at other times in its development (Kolb and Nonneman, 1975). Seeing the kitten on more than one occasion can therefore provide a more accurate picture of its personality and temperament.

Fig. 9.2. If you have a dog, or are considering getting a dog, it is advisable that your new kitten has had regular positive experience with dogs. But because of the tremendous variation between different dog breeds and types, the effects of positive socialization might be limited to dogs of a similar size and breed type as those it has become accustomed to at an early age.

At what age should a kitten leave its mother to go to its new home?

The Governing Council of the Cat Fancy (GCCF) recommends that kittens should not go to their new homes until they are at least 12–13 weeks of age, and a recent study suggests that 14 weeks may be the optimum minimum age for homing, to improve the kittens' welfare and prevent future behaviour problems such as aggression and stereotypic behaviours (Ahola *et al.*, 2017). However, many non-pedigree kittens are sold or passed on to their new homes when they are around 8 weeks of age, or sometimes younger, around 6–7 weeks of age. Kittens may be able to eat solid food at 6–8 weeks of age but are unlikely at such a young age to be fully weaned, and are less likely to be emotionally prepared to leave the nest, which can lead to future behaviour problems.

If acquiring an adult cat or older kitten

- Try to find out as much as possible about the cat's past. A cat or kitten from a feral background, or one where it has received very little positive socialization with people and habituation to family life, is less likely to become a suitable pet. The cat may eventually become accustomed to, and even become closely bonded with, its owner and close members of the household but it is very likely to remain fearful of all other people. Also, life as a pet in a normal family home can be highly stressful for such cats.
- Make sure that you are able to spend time interacting and handling the cat or kitten and that it appears confident and friendly, or recovers fairly quickly from any initial fear or trepidation.
- If taking on an adult cat, ensure that it is neutered and fully vaccinated and free from disease and parasites, especially if you have other cats at home.

Bringing your New Cat or Kitten Home

Being taken to a new home, away from the comfortable security of mother, littermates and familiar surroundings, can be a highly stressful time for a kitten. An adult cat can also experience stress when being taken away from a home to which it has become accustomed. Making the transition as easy as possible for the new cat or kitten is therefore an important factor in helping your new pet settle in.

- A few days before bringing your new cat or kitten home, take a cloth containing the scent of your home to the place where the new cat or kitten is resident, e.g. the breeder's home or rescue centre. Ask for this to be placed with your new cat to allow him to become accustomed to the scent of his new home.

Preparation of a 'safe room'

Prepare a 'safe' room for the new cat or kitten. This should be somewhere quiet, away from other pets (especially other cats), children, loud noises and lively activity.

This room should contain all that the cat or kitten needs:

- Food.
- Water – placed away from the food.
- Litter trays – placed away from food and water.
- Comfortable and warm bedding.
- Toys.
- Hiding and 'safe' places – for example, cardboard boxes, access under furniture and/or high places.

The safe room should also be somewhere that the cat or kitten will later have easy access to.

The homecoming

- Always transport the cat or kitten in a secure purpose-built cat carrier. Hold the carrier close to your chest rather than by the handle, as the 'swinging' motion when you walk may be disorientating and unpleasant for the cat or kitten.
- A cloth or item of bedding containing scent from the cat or kitten's current home should go into the carrier as well.
- If transporting by car, the carrier should be well secured using a seat belt or similar to prevent it from moving around.
- When you reach your home take the cat or kitten in the carrier straight to the pre-prepared 'safe room'.
- Open the carrier and allow the cat to exit in his own time. Don't lift him out or make any fuss.
- Allow the new cat or kitten to explore the room and/or hide as he wants. Don't try to restrain him or remove him from a hiding place. Give him time.
- Keep the carrier containing the 'familiar scent cloth' in the room with him. Or if he finds a preferred hiding place put the cloth there.
- Allow the new cat or kitten to become settled and relaxed in this safe room before allowing him access to the rest of the house.
- If you have other cats, see Appendix 5 for further advice. If you have a dog, see Appendix 7.

References

Ahola, M.K., Vapalahti, K. and Lohi, H. (2017) Early weaning increases aggression and stereotypic behaviour in cats. *Scientific Reports* 7(1) DOI: 10.1038/s41598-017-11173-5.

Bernstein, P.L. (2007) The human–cat relationship. In: Rochlitz, I. (ed.) *The Welfare of Cats*. Springer, Dordrecht, The Netherlands.

Caney, S. and Halls, V. (2016) Feline bereavement. In: *Caring for an Elderly Cat*. Vet Professionals, Edinburgh, UK, pp. 113–117.

Dinis, F.A. and Martins, T.L.F. (2016) Does cat attachment have an effect on human health? A comparison between owners and volunteers. *Pet Behaviour Science* 1, 1–12.

Finkler, H., Gunther, I. and Terkel, J. (2011) Behavioural differences between urban feeding groups of neutered and sexually intact free-roaming cats following a trap-neuter-return procedure. *Journal of the American Veterinary Medical Association* 238, 1141–1149.

Friedmann, E. and Thomas, S.A. (1995) Pet ownership, social support and one year survival after acute myocardial infarction in the Cardiac Arrhythmia Suppression Trial (CAST). *American Journal of Cardiology* 76, 1213–1217.

Kanat-Maymon, Y., Antebi, A. and Zilcha-Mano, S. (2016) Basic psychological need fulfilment in human–pet relationships and well-being. *Personality and Individual Differences* 92, 69–73.

Karsh, E.B. and Turner, D.C. (1988) The human-cat relationship. In: Turner, D.C. and Bateson, P. (eds) *The Domestic Cat: The Biology of its Behaviour*, 1st edn. Cambridge University Press, Cambridge, UK.

Kessler, M.R. and Turner, D.C. (1999) Socialization and stress in cats (*Felis silvestris catus*) housed singly and in groups in animal shelters. *Animal Welfare* 8, 15–26.

Knudsen, E.I. (2004) Sensitive periods in the development of the brain and behaviour. *Journal of Cognitive Neuroscience* 16, 1412–1425.

Kolb, B. and Nonneman, A.J. (1975) The development of social responsiveness in kittens. *Animal Behaviour* 23, 368–374.

10 Advice for Cat Owners

Early influences on development such as genetic factors, exposure to early life stressors and experiences during the sensitive period (2–7 weeks of age) can certainly increase or decrease the likelihood of behaviour problems developing later in life (see Chapters 5 and 8). But of no less importance are the influences that the cat is exposed to as an adult. Fear or anxiety owing to real or perceived danger, or frustration because of competition for important resources or insufficient opportunity to express normal behaviours are just some of the factors that can increase stress and increase the probability that the cat may start to exhibit behaviours that are not only unpleasant for the owner, but may also be indicative that the cat is suffering poor behavioural and emotional welfare.

Managing Feline Stress

Feline stress is a major cause of both behavioural and health issues (see Chapter 6). Being aware of the most likely causes of stress for pet cats and managing care and the cat's environment to minimize stress are therefore fundamental to feline welfare, reducing the risk of developing behaviour problems.

Among the most common causes of stress for pet cats are:

- A lack or insufficient opportunity to engage in natural behaviour.
- Perceived insufficient resources or competition for resources, usually with other cats.
- Conflict with other cats.
- Abusive handling.
- Inconsistent or unpredictable handing or husbandry routines.
- Poor health.

Indoor Cat or Outdoor Access?

There is much debate as to whether pet cats should be allowed to venture outside or be kept confined to the safety of the home. There is also notable cultural division in the general decision to allow cats the freedom to roam. In the UK it is not uncommon for a pet cat to be allowed the opportunity to go outside and to roam wherever it pleases for at least some of the day. But in other areas of the world it is more customary for pet cats to be confined indoors (Rochlitz, 2005).

The main advantage of unrestricted outdoor access is that it provides the cat with the best opportunity to engage in natural behaviours such as exploring, hunting and predatory play. It can also allow a cat to escape and spend time away from a household where there may be constant stressors such as conflict and competition with other household cats.

But the disadvantage is that it can expose the cat to increased physical danger from traffic, predation by other animals, fights with other cats and increased disease risk. Risks to local wildlife from feline predation is another important consideration and in some areas there are regulations restricting outdoor access for pet cats (Grayson and Calver, 2004). If your cat is allowed free outdoor access, teaching him to come to call can be very useful (see Appendix 9).

Cats that are confined indoors are kept safe from the dangers they might be exposed to outside, but decreased opportunity to engage in natural behaviours can be detrimental to behavioural welfare and the emergence of behaviour problems can be greater with indoor cats if they are not provided with sufficient space and environmental enrichment (see Appendices 1 and 2) (Heidenberger, 1997; Strickler and Shull, 2014).

Semi-confinement can provide a compromise between allowing the cat free outdoor access and total indoor confinement.

A Fenced-off Area

This can be an excellent way of keeping your cat safe while still allowing him to enjoy outdoor life.

Cat-proof fencing

Because cats are small, agile creatures able to climb, jump and squeeze through small gaps, fencing your entire garden or an area of your land to securely confine your cat will usually require careful design and consideration. Specialist fencing, designed to prevent cats from jumping or climbing over it, either on its own or added to current walls or fences, can often work very well (Fig. 10.1).

Electric containment systems

This works by way of a buried wire that sends a signal to a collar worn by the cat. When the cat approaches the wire a warning sound is emitted and if the cat continues in the same direction it will receive a 'correction' via the collar that is said to be similar to static electricity. This physical correction can usually be deactivated after a time because the cat learns to avoid the areas where it has previously received the unpleasant stimulus and the sound alone becomes sufficient to prevent the cat from crossing the invisible barrier. Although this system is popular with some who report that it works well for them, it is important to be aware of the potential detrimental issues that can be associated with it:

Fig. 10.1. Fencing your garden can work well but may require specialist fencing. Photo courtesy of Protectapet (www.protectapet.com).

- It provides no deterrent or barrier to rival cats or predators. But fear of crossing the barrier may prevent the resident cat from escaping to safety.
- Although the 'correction' is designed to be mild, pain is a subjective experience and some cats may need a stronger shock than others to effectively prevent them from crossing the barrier.
- The cat's motivation to cross the barrier is another factor. If the cat is frightened by something or being chased or is highly motivated to chase prey, it may ignore the warning signal and even the physical correction. However, if it is not so motivated to make the return journey, the need to avoid the potential shock may prevent it from getting back home.
- Cats can vary regarding their sensitivity to negative experiences and what they learn from them. For some the experience of the correction and the sound associated

with it can be highly stressful, which might result in generalized, and possibly severe, fear and anxiety.
- It is also very important to be aware that the use of these systems is illegal in some countries and under review in others.

A purpose-built enclosure

If fencing a large area is not possible, another option is to erect a purpose-built enclosure (Fig. 10.2). The structure can be built attached to the house, allowing the cat to access it directly via a cat flap or door, or built at a distance from the house, which the cat can either access via a run or be transported there and back by the owner.

Fig. 10.2. An example of a purpose-built cat enclosure. Photo courtesy of Chris Stalker.

Particularly if the cat does not have constant access back into the house whenever it wants, an outdoor enclosure should contain all that the cat requires, including:

- Food and water.
- Litter trays.
- Comfortable resting areas.
- Shelter from cold and wet and access to both sun and shade.
- Multiple hiding places and elevated areas.
- Appropriate environmental enrichment (see Appendix 1).

More detailed information on cat-proof fencing and enclosures can be accessed from International Cat Care (https://icatcare.org/advice/fencing-your-garden).

Lead and harness

Some cats adapt well to being taken for a walk on a harness and lead, and this can also be a good way to introduce a young cat to the great outdoors and allow it to explore its territory before allowing it out by itself. However, not all cats will accept wearing a harness and some can become frustrated from being restrained or increasingly fearful and stressed if they are prevented from escaping a real or perceived threat encountered on a walk.

Introducing a cat to wearing a harness can have a greater chance of success if started when the cat is young, preferably under 1 year old, and should be approached gradually, using praise and food treats to encourage good associations with the harness. It is important also to ensure that the harness is comfortable and well-fitting, and that the cat is relaxed when wearing the harness, before attaching a lead and taking the cat outside.

Cat Flaps

Being reliant on a human to open doors to allow access outside or back inside away from bad weather and threats such as other cats, can be a source of stress and frustration for a pet cat. Fitting a cat flap to allow the cat the freedom to come and go as he pleases can be an ideal solution. However, a cat flap can also be a source of stress and conflict with other cats if it allows neighbouring cats access into the home or if it is perceived by the resident cat(s) as being a possible access point for intruders. Where the cat flap is positioned and the type of cat flap fitted can, however, make this much less of an issue (see Appendix 8).

Neutering

Neutering (also see Appendix 4) is essential for pet cats that are allowed free outdoor access, but it can also be an important part of stress limitation and reducing unwanted behaviours such as urine spraying and aggression in both outdoor and indoor cats. Unneutered cats are far more likely to roam further and get into serious fights with other cats. Sexually entire females are also very likely to become pregnant if allowed to roam, and it is important to be aware that, although 6 months or older is the usual age for a female cat to come into season for the first time, they can reach puberty as young as 4 months of age (Joyce and Yates, 2011).

The strong desire to find a mate or to fight with rivals can make a cat less aware of its surroundings and therefore be more at risk from road traffic accidents and other physical dangers. Fighting also exposes the cat to a much greater disease and injury risk. Conflicts can still occur between cats that have been neutered but the frequency and physical intensity of fighting, and so the risk of serious injury, is much reduced.

Other Cats

Conflict or confrontations with other cats, either with known cats within the home or unknown cats outside, is a common stressor resulting in health issues and behaviours such as fighting, anxiety and indoor urine marking.

Multi-cat households

Close social bonds can and do occur between cats, and pet cats that are in a close, friendly relationship with one or more other cats do appear to enjoy and benefit from it. However, not all cats that live together get along well and even when the relationship may appear amicable, tension and competition may exist between them and be a major source of stress (see Appendix 16). Plus, even close bonded relationships between cats can be easily broken and may not be so easily repaired.

There are many factors that can influence the nature and quality of a relationship between cats in a multi-cat household and it is advisable to be aware of these to maintain peace in a household with more than one cat.

The most common cause of stress and conflict between cats is competition for resources such as food, water, resting places and even litter trays. Appendix 3 provides information on how resource competition between cats that live in the same household can be managed.

Adding another cat to the household can also cause major upset and potential long-term stress, unless given plenty of consideration beforehand, and carried out in a manner that aims to minimize stress and conflict for both the resident cat(s) and the incomer (see Appendix 5).

Neighbouring cats

Being allowed outdoor access can be highly beneficial in that it can be the easiest and most effective way of providing environmental enrichment and allowing a cat to perform natural behaviours. But there are also many things that a cat may find frightening outside, such as encounters with dogs, unwelcome attention from people and loud or sudden noises (e.g. thunder, fireworks, etc.). But the most common stressor is other cats in the neighbourhood. There are, however, ways in which your garden and the area close to your home may be made a more secure place for your cat.

- Strategically placed bushes and solid fencing may help to deter other cats from entering your garden, but completely preventing other cats from getting onto your land is difficult. Cat-proof fencing can be used, but unless the area is totally enclosed (Fig. 10.3) fencing will not completely prevent other cats from getting in and once in the area they will not be able to get out. It is anecdotally reported, however, that neighbouring cats are much less likely to return if they have entered the area and then not been able to get out again until released. Fencing can be erected to keep other cats out, but because this can only be effective by having an overhang on your neighbour's side of the fence or having alterations made to their side of the fence, this can only be achieved with the neighbour's permission.
- A major source of stress for resident cats is having neighbouring cats looking down on them from elevated positions. Preventing or discouraging other cats

Fig. 10.3. If an area is completely enclosed with a roof, this will also prevent other cats from entering. Photo courtesy of Protectapet (www.protectapet.com).

from sitting on the tops of walls, fences or other tall structures can therefore help your cat to feel less threatened. This may be achieved by placing pot plants or similar along the tops of walls or fences or by erecting wooden trellis or more purpose-made 'cat deterrent' fencing along the top of the wall or fence; however, do not use anything that may potentially cause pain or injury.

- Also avoid terracing in the garden, especially around the periphery because this can provide another area for your cat's rivals to position themselves in an elevated area and so be more of a threat to your cat.
- Using evergreen bushy plants and large items such as garden furniture, ornaments, outdoor cat trees, or even children's play structures can provide places in the garden for the cat to hide and/or get up high, and so help to increase his sense of security.
- Placing low bushy plants near to the outside of your cat's regular exit point from the house can also provide an area from where he can hide and assess his surroundings before venturing further.

Real or perceived threats from outside when your cat is indoors

Even exclusively indoor cats can feel threatened by neighbouring cats if the other cats are able to enter the house or can be seen through windows or glass doors (Fig. 10.4).

Fig. 10.4. Seeing other cats through a window can also be stressful for many cats.

- If a cat flap is installed, ensure that it is an exclusive type that can only be used by your cat(s) (see Appendix 8).
- Position food bowls, litter trays and resting areas well away, and not within sight of cat flaps, doors or windows through which other cats may be seen.
- If other cats can be seen through glass doors or windows, block your cat's view to outside by using pot plants placed by the window or by covering the area that the cat can see through (usually the lower third to half) with a temporary window frosting.

Cats, Babies and Children

A new baby or children in the house, especially if the cat is not accustomed to babies or children, can be another major source of feline stress. Situations may also arise where the cat feels threatened, which could result in the child being bitten or scratched. Advice on how to keep babies and young children safe and how to minimize stress for your cat can be found in Appendix 10.

Cats and Dogs

The traditional view is that cats and dogs are sworn enemies, but it is very possible for cats and dogs to live together in harmony. However, it should always be kept in mind that dogs are natural predators and, although cats are also predators, they are also small enough to be regarded as potential prey by a dog. Making the right choices and taking time to introduce cat and dogs together correctly (see Appendix 7) is important. Of no less importance is continued care and consideration for the cat's welfare to ensure that the cat is not put in danger or subjected to an enduring stressful existence when living in the same house as a dog.

- Ensure that the cat always has regions within the house where he can get away from the dog. This may involve using baby gates or internal cat flaps to section off 'cat only' rooms or areas of the house.
- Ensure that there are plenty of elevated areas, e.g. high furniture, shelves, work surfaces or tabletops available to allow the cat to jump up and be out of reach of the dog.
- Ensure that the cat has access to resting areas away from the dog.
- Ensure that your dog is well trained and under control so that you can call him away from the cat when necessary.
- Be aware of your cat's body language (see Chapter 3) so that you can see if your cat is feeling threatened by the dog.

Avoiding House-training Issues

Urinating or defecating in the house away from the litter tray is a common feline behaviour problem. Advice on how best to avoid or deal with this issue can be found in Appendix 11.

Preventing Human-directed Aggression

Cats are armed with a mouthful of sharp teeth and four feet with sharp claws, and they can cause a fair amount of injury if they direct these 'weapons' towards us. It is therefore important to be aware of why a cat might become aggressive and what we can do to prevent human-directed aggression.

Predatory-type play

Cats are natural hunters and a large part of feline play involves practising predatory skills. Although fun for the cat, when directed towards hands, feet or any other human body part this type of play is not so much fun for us.

- Do not encourage your cat or kitten to play with your fingers or toes, etc. It can be hard to resist wiggling your fingers to get a little kitten to chase and 'attack', but this teaches the kitten that people can be used as 'toys'. Also as the kitten grows up and becomes stronger, more force will be applied making this type of 'play' more painful for the victim.
- Do encourage appropriate play with toys (see Appendix 2).
- If the cat or kitten makes a 'playful' attack on you: (i) because feline predatory behaviour is triggered and encouraged by movement, it is best to try to keep as still as possible; (ii) re-direct the cat or kitten towards a moving toy; and (iii) do not attempt to punish the cat or kitten because this may increase the aggression by causing the cat to become defensive.

Defensive aggression

Defensive aggression occurs when the cat experiences pain or fear, or anticipates that something painful or frightening is about to happen. Minimizing stress, fear and pain as much as possible is therefore important to reduce the risk of human-directed aggression.

A cat that has had a good start in life, been well socialized and had mainly positive experiences with people, both as a kitten and later in life, is less likely to feel fearful of people or become aggressive towards them (see Chapters 8 and 9). But it is also important that positive experiences with people continue throughout the cat's life.

- Do not attempt to punish the cat physically or in any way that might cause the cat to feel frightened.
- Cats usually cope with frightening events by running away and hiding, or by jumping up onto an elevated position from where they can look down on a perceived threat. If they are unable to do any of these, the chances are increased that the cat may resort to aggression as a defence. Even if you are not the main source of fear, cats may re-direct aggression onto whatever or whoever is nearest if they do not have other means of coping. It is therefore important that pet cats are provided with plenty of escape routes, places to hide and access to elevated spaces such as the tops of furniture, shelves, tables and work surfaces.
- Because of the risk of re-directed aggression, avoid restraining your cat or attempting to reassure him by picking him up if he is feeling frightened or agitated.

Keeping Your Cat Healthy

There is a strong link between behaviour and disease (see Chapter 6). A change from normal behaviour can be one of the first signs that a cat is unwell; also cats that are in pain or poor health are more likely to experience stress and exhibit behaviour problems. It is therefore important for your pet cat's overall welfare to keep him in good health.

- Feed a healthy, well-balanced, meat-based diet. It is best to take the advice of your vet or vet nurse/technician as to the best diet for your cat.
- Keep up to date with vaccinations, worm and flea treatments. It is advisable to use products recommended by your vet. Also ensure that you only use products that are safe for cats, and for the age and size of your cat.
- Ensure that your cat is given a regular veterinary check-up at least once a year.

One reason why a cat may not be taken to see the vet for regular check-ups, or even when showing signs of illness, is the difficulty that can sometimes be involved. The first obstacle can be getting the cat into the cat carrier and transporting it to the veterinary practice. This can be dealt with by training the cat to regard the carrier as a place of comfort and safety, rather than as something to be feared and avoided (see Appendix 12). The other problem can be the cat's fear while at the vet which can intensify at each visit and not only be highly stressful for the cat but also cause the cat to become increasingly difficult to handle and examine.

The International Society of Feline Medicine (ISFM) has set up a worldwide programme that aims to help veterinary practices to become 'Cat Friendly Clinics' where the sources of stress and fear for cats associated with veterinary visits are reduced, which can make visits to the vet much easier for both cats and their owners. In the USA, the programme, known as 'Cat Friendly Practice' is licensed by the American Association of Feline Practitioners (AAFP) (for further details see: www.catfriendlyclinic.org in the UK and countries other than the USA, and www.catvets.com/cfp/cfp in the Americas).

References

Grayson, J. and Calver, M.C. (2004) Regulation of domestic cat ownership to protect urban wildlife: a justification based on the precautionary principle. In: Lunney, D. and Burgin, S. (eds) *Urban Wildlife: More than Meets the Eye*. Mosman, Royal Zoological Society of New South Wales, Australia, pp. 169–178.

Heidenberger, E. (1997) Housing conditions and behavioural problems of indoor cats as assessed by their owners. *Applied Animal Behaviour Science* 52, 345–364.

Joyce, A. and Yates, D. (2011) Help stop teenage pregnancy! Early-age neutering in cats. *Journal of Feline Medicine and Surgery* 13, 3–10.

Rochlitz, I. (2005) A review of the housing requirements of domestic cats (*Felis silvestris catus*) kept in the home. *Applied Animal Behavior Science* 93, 97–109.

Strickler, B.L. and Shull, E.A. (2014) An owner survey of toys, activities, and behavior problems indoor cats. *Journal of Veterinary Behavior: Clinical Applications and Research* 9, 207–214.

11a Advice for Veterinary Professionals
Part 1 – The Cat in the Veterinary Clinic

Knowledge and expertise in feline medicine and surgery has grown tremendously during recent years and general recognition of the importance of feline health care has also increased in the cat-owning population. Even so, it is not uncommon for even the most caring and devoted cat owner to be reluctant to take their pet to the veterinary surgery for preventative health care, or even for treatment when their pet is injured or becomes ill.

This reluctance is most often due to one or more of the following (Vogt *et al.*, 2010):

- Challenges in transporting the cat to the practice:
  - Difficulty in getting the cat into the cat carrier, which may even result in injury to the owner or handler.
  - Obvious distress exhibited by the cat while travelling.
- Difficulty in handling the cat at the practice:
  - Owners may feel embarrassed or fear that they will judged as inept pet carers if their cat is aggressive or difficult to examine.
  - They might worry that their cat will cause injury to veterinary staff or to themselves.
  - They may find themselves facing extra expense if the cat needs to be sedated in order to treat or examine it.
- Unwanted changes to the cat's behaviour following a veterinary visit:
  - A cat may act aggressively towards the owner and/or towards other household pets following a visit to the vets.
  - The relationship between feline housemates may be severely and sometimes irreparably damaged following a veterinary visit.
- Empathy for the cat:
  - An owner may feel that the possible drawbacks and the distress caused to the cat by a visit to the vet outweighs the health benefits (Habacher *et al.*, 2010).
- Stress for the cat owner:
  - As well as being stressful for the cat, the experience can also be sufficiently stressful for the owner that they may choose to simply avoid future visits or go elsewhere if they feel that another practice may be more 'feline friendly' (Cannon and Rodan, 2016a).

Endeavouring to make life easier for clients and increasing the likelihood that they will seek veterinary care for their pets is therefore important, but this is not the only reason why it is important to minimize fear and stress for feline patients.

- Cats that experience fear and stress are more likely to exhibit fear-based defensive aggression and become increasingly difficult to handle.
- Stress may delay healing and recovery from disease and injury (Padgett and Glaser, 2003; Gouin and Kiecolt-Glaser, 2011).
- Stress can significantly alter the results of physical examinations and laboratory tests, leading to incorrect diagnoses and treatment (see Box 11.1).
- Stress can have many significant negative effects on feline health and welfare (see Chapter 6).

The Journey to the Practice

Minimizing the cat's stress associated with the journey to the practice can also help to reduce stress and reactive behaviour at the surgery. Practice staff can offer advice and guidance as to how travelling may be made easier for cats and their owners and how associated stress may be reduced (see Appendix 12).

The Waiting Room

The sights, sounds and smells of other animals and stimuli associated with the veterinary practice can make the waiting room a frightening and highly stressful place for a cat; however, there are ways in which these stressors may be significantly lessened:

- Provide a separate cat-only waiting area:
  - Provide a separate room for feline patients or arrange separate consultation times for cats and dogs. An unused consultation room could be utilized as a cat-only waiting room.
  - If this is not possible, cordon off an area of the waiting room as a designated cat-only area.
  - If possible, the cat waiting area should be a significant distance away from the dog area, and should be positioned so that it is not necessary for dogs to be walked through the cat area for any reason. If space is limited, solid barriers should be erected so that dogs cannot be seen, even if it is not possible to prevent them from being heard or smelt.

Box 11.1. Physiological and diagnostic parameters influenced by stress.

- Changes to heart rate – usually increased but can be decreased if stress is chronic.
- Increased respiratory rate.
- Alkaluria (increased urine pH owing to increased respiratory rate).
- Increased rectal temperature.
- Dilated pupils.
- Hypertension (raised blood pressure).
- Stress hyperglycaemia (raised blood glucose).
- Raised white blood cell count if stress is chronic.
- Diarrhoea or stress colitis.

(Cannon and Rodan, 2016a; Sparkes et al., 2016)

- Provide shelves for cat carriers:
  - Cats can feel more at threat when at ground level, especially whilst confined in a cat carrier and therefore unable to escape from perceived threats. Being in an elevated position can make them feel more in control and less threatened. However, care must be taken that if a cat in a carrier is placed on one of the shelves provided it does not overlook dogs in another area of the waiting room.
  - Places to put cat carriers above 'dog level' should be provided at the reception desk because this is one place where cats and dogs may still meet even if separate waiting areas are provided.
- Provide towels or similar that clients can use to cover their cat carriers:
  - Part of a cat's natural defence mechanism is to hide in a darkened area. Covering the carrier can go a long way in helping the cat to cope, and help to minimize fear and stress. This can be especially important at the reception desk where the cat is more likely to be exposed to stressors such as the noise of telephones, the sight of other animals and strangers peering into the carrier.
- Provide something for clients, such as a waiting room television and/or interesting and informative posters, magazines or leaflets to read:
  - The owners' behaviour and attitude may also influence that of their pet. Providing interesting distractions can help to reduce stress for the pet owner, especially if they have a long wait or if they are concerned for their pet.
  - This can also help to reduce the risk of other pet owners peering into the cat carrier and attempting to 'make friends' with the cat. Although usually intended as a kindly gesture, this can be frightening for some cats, especially as the cat is confined in the carrier and unable to escape.
- Inform clients:
  - Providing separate areas, shelves and covers are a good idea but of little use if clients are unaware of them or why they have been provided. Reception staff may not always be able or have the time to explain these things to clients, so clearly visible notices should be put up for client information and to encourage use of these facilities.

Appointments

It is far better to make set appointments than to have open surgeries, which can result in a crowded waiting room and long waiting times. The length of each appointment should also allow sufficient time for the cat to be allowed to settle if necessary, and for gentle, calm and unhurried handling to take place.

Greeting and Speaking to Clients

All members of practice staff should have good communication skills and endeavour to greet and speak to clients in a kind, compassionate, empathic and friendly manner. Friendly greetings and interactions at each visit can help to build client bonds and the client's trust in the practice.

It should also be recognized that owners might be in a state of distress, possibly owing to difficulties they have experienced in getting the cat to the practice, anxiety about the possible outcome of the veterinary visitation for a beloved pet, or having just had bad news about their pet's health. Although no member of practice staff should be expected to be a victim of abuse or threatening conduct, it is important to be aware that stress and anxiety experienced by an owner may influence their behaviour, causing them to be less patient or tolerant and less able or willing to communicate their concerns, answer essential questions, or to acknowledge and understand advice they are given. Stress might even influence their interactions with their pet.

Endeavouring to minimize stress for the client as well as the patient is therefore an important part of overall care.

The Consultation Room

- Because cats may feel threatened just by the scent of dogs, it can be better to use a designated 'cat-only' consultation room.
- Some cats will relax more easily if they are permitted to explore any new environment to which they are exposed. Consulting rooms therefore need to be escape-proof and devoid of gaps and spaces where a cat may become trapped or could have access to anything that could pose a possible injury risk.
- As well as a consulting table, chairs should also be provided, not only for the clients' use but also to provide an alternative area to examine feline patients that might feel more relaxed and amenable to examination when sat on the practitioner's or owner's lap.
- The consultation room must be large enough to contain the examination table, chairs and all necessary equipment whilst still allowing sufficient comfortable space for the vet, a nurse/technician and the clients.
- The consultation table should be covered in a soft, non-slip, and easy to clean material.
- To reduce the need for the vet or nurse having to repeatedly leave and re-enter the room, any veterinary or handling equipment likely to be needed on a regular basis must be readily at hand within the consulting room.
- Avoid automatic air fresheners. Not only can the scent be overpowering for cats but such air fresheners often make a hissing sound that can be frightening for them.

Handling and Examining

Minimal, gentle restraint and handling should always be employed (Fig. 11.1). Avoid 'scruffing' (grabbing and holding tight the loose skin at the back of the neck) (Fig. 11.2) or other harsh means of restraint. These practices, which can cause pain, discomfort and fear, might appear to be successful means of restraint, but the long-term effect can be to increase the cat's stress and result in the increase or development of defensive aggression towards handlers in the future.

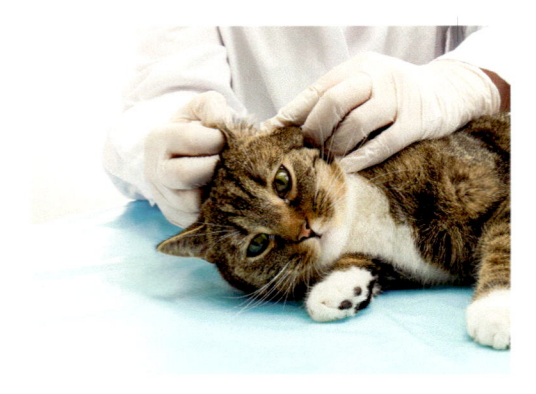

Fig. 11.1. Minimal, gentle restraint and handling should always be employed. If a cat is difficult to handle it can be wrapped in a towel.

Fig. 11.2. Never scruff a cat unless it is absolutely unavoidable. This type of handing can be uncomfortable and frightening for the cat and can increase the likelihood that the cat might become aggressive.

Handling

Removing the cat from the carrier:

- **Do not** attempt to tip or shake the cat out of the carrier, physically pull it out, or scruff the cat to extract it.
- Allow the cat to exit the carrier in its own time or encourage it out with a food treat. Some cats can feel more at ease and be more willing to leave the carrier if it is placed on the ground.
- If the carrier can be separated into two halves, remove the top half so that the cat can be easily lifted out from the bottom half.
- If there is bedding in the carrier, lifting this out with the cat can help it to feel more secure.
- If necessary, approach from behind and wrap the cat gently in a towel and then lift it out.
- Once the cat has been removed from the carrier, place the carrier away from the cat, preferably out of the cat's direct line of vision. If the carrier is nearby and

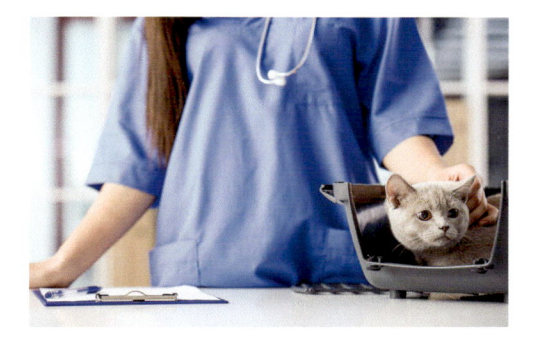

Fig. 11.3. A cat may be more relaxed and easier to examine if allowed to remain in the carrier.

within the cat's sight, it may attempt to retreat into the safety of the carrier and become agitated if prevented from doing so.
- If the carrier has a large top opening or can be separated into two halves, it can be easier and less stressful for the cat if it is examined in the carrier (Fig. 11.3). If the cat is fearful and defensive, use a towel to cover it and only expose the area that needs to be examined.

Examination

- Start with the cat facing away from you.
- Stroke the cat, especially the areas of the facial skin glands, i.e. under the chin, the corners of the mouth and either side of the forehead (see Chapter 3).
- Talk to the cat in a calm and measured voice, and keep to the same cadence and tone of voice when speaking to the clients.
- Avoid making 'shhh' sounds or similar that the cat may misinterpret as hissing.
- Avoid fast, sudden or unpredictable movements.
- Avoid prolonged direct eye contact.
- Always start with the least painful area and progress towards areas that are more likely to be painful or uncomfortable.
- If the cat is difficult to examine, wrap the cat gently yet sufficiently, but not over-tightly, in a large towel so that only the area that requires examination is exposed.
- Not all cats will take food treats, but for those that will there can be an increased chance that the cat may learn a positive association with handling and with the veterinary context. Having a variety of highly palatable treats available can therefore be a good idea, especially for kittens, which are more likely to take food treats and less likely to have already made a negative association with the clinic.
- Don't restrict examinations to the examination table. Other places where a cat may feel less stressed and may be easier to examine can include:
  - On the owner's lap. The practitioner should be on the same level, and not leaning over the cat.
  - On the practitioner's lap.
  - On a chair or bench while the practitioner, and possibly the owner as well, is sat next to the cat.
  - On the floor.
  - On an elevated shelf.

- After the examination and/or treatment allow the cat to go back into the carrier of its own accord.

Compiled with reference to Cannon and Rodan, 2016b.

Hospitalization

Recognizing stress in hospitalized cats

Recognition of stress in a hospitalized cat can be difficult because the same behaviours can also be due to pain, discomfort and simply feeling unwell. However, as pain and sickness are sources of stress in themselves, it may be reasonable to assume that hospitalized cats exhibiting any of the following are likely to be suffering from stress:

- Decreased activity.
- Reduced appetite.
- Reduced sleep or sham sleeping (pretending to sleep).
- Lack of self-grooming or excessive grooming.
- No interest in play or affection.
- Hiding or attempting to hide.
- Retreating to the back of the cage.

Fearful cats may also demonstrate any of the following defensive behaviours and signs of increased arousal, especially when approached and attempts are made to handle or interact with them:

- Dilated pupils.
- Ears flattened back or to the side.
- Increased vigilance.
- Piloerection.
- Vocalizing (growling, shrieking).
- Aggression.

As well as fear and anxiety, frustration can also be a component of stress, so any of the following signs of frustration in caged cats (Fig. 11.4) can also be a cause for concern:

- Excessive vocalization.
- Continuous or frequent reaching and grabbing through the bars of a cage.
- Destructive behaviour (ripping up bedding, tipping up food, water, and/or litter trays) (Gourkow et al., 2014).

Because of the negative and harmful effects of stress on disease and recovery, it is important to try to limit stress as much as possible for hospitalized cats.

Housing

- A separate cat ward is preferable, away from the sight and sound of dogs and located so that it is not necessary for dogs to be walked through the cat ward, or for cats to be carried through the dog ward in order to access other areas.

Fig. 11.4. Frustration is also an aspect of stress in hospitalized cats.

- The cat ward should also be isolated from other intrusive sounds and activity, such as telephones, alarms, the sound of metallic instruments or equipment being used or cleaned, and noisy machinery such as washing machines, etc.
- Cat cages should also be designed or adapted so that they can be opened and closed with minimal noise and/or reverberation.
- Avoid bright or harsh lighting and avoid or cover reflective surfaces within the cages.
- The ward should be escape-proof with a self-closing door, and without gaps or spaces where a cat may become hurt or trapped.
- The ward should be kept at a constantly warm (20–25°C) but not overly hot ambient temperature. If individual heating is provided, e.g. lamps or heat pads, ensure that they are safe and do not become over hot and increase the temperature to an uncomfortable level for the patient.
- Cages should be large enough to contain the following:
  - Space for food, water, comfortable bedding and a litter tray to be positioned with as much distance as possible between them.
  - An elevated area such as a shelf or solid structure that the cat can climb on.
  - A hiding place, e.g. a cardboard box, a solid-sided cat carrier, an igloo bed or high-sided bed, a covered litter tray converted into a bed, or a towel or similar draped over a fitted shelf to provide a hiding place. All these items should be easy to remove to allow for cleaning and for any time when the cat may be at risk of harm from obstructions within its hospital cage, for example, during recovery from anaesthetic.
- Cages should be sufficiently large but not be so deep that removal of the cat from the cage is difficult.
- It is preferable that cages are situated so that hospitalized cats cannot see each other. If this is not possible a towel can be used to cover the cage door if a cat appears fearful, anxious or requires increased sensory isolation.
- Cages should not be at ground level or positioned so high that it is difficult to access or observe the cat. Not only can this make it difficult and potentially dangerous should the cat become fractious when attempts are made to tend to it or remove it from the cage, it can also make it very difficult to observe the patient.
- Avoid using any strong smelling disinfectants, cleaners or air fresheners.

Nursing care of hospitalized cats

- Unpredictability can be a major stressor for cats, so aim to keep to a reliable daily routine:
    - Cage cleaning, feeding, grooming and general care of the cat should be carried out at the same time each day and preferably by the same person.
    - Daily treatment and examination should also be conducted at a consistent time, and preferably by someone other than the person conducting daily care.
    - Owners should also be requested to visit at the same time each day.
- Using bedding from the cat's home that contains familiar scent can help to increase the cat's sense of security. Ask the owners to bring in bedding or use the bedding contained in the cat's carrier if suitable.
- Owners can also be questioned as to the cat's preferred food and litter substrate, especially for long-term hospitalization. Owners can also be requested to bring in some of the cat's favourite food and stay with the cat while it eats, especially if the cat is otherwise reluctant to eat.
- It can be highly stressful for cats to be observed, or to observe other cats undergoing veterinary examination or treatment (Wallinder *et al.*, 2012). Therefore, do not examine or treat cats within sight of other cats, or other animals, or allow them to witness other cats undergoing examination or invasive procedures.
- Being watched or approached by other unknown cats whilst confined in a cage can be a major stressor. For this reason, cats should not be permitted to roam free within the hospital ward if the cat cages are on ground level or if free roaming cats are able to approach the caged cats.
- Gentle, calm handling and restraint as described previously should be continued.
- Adequate analgesia should be administered to minimize acute and chronic pain.

Cleaning the cage

Keep the cat in the same cage during the period of hospitalization. The familiarity of the hospital cage can be the best thing that the cat has as a place of safety in an unfamiliar and hostile environment. To maintain this, it is best that the cat remains in the same cage throughout the period of hospitalization. Frequent re-location from one cage to another, for cleaning or other purposes, should be avoided, and cleaning should be conducted with the cat in the cage as follows:

- Approach calmly while speaking to the cat in a relaxed, friendly voice.
- Open the door quietly.
- Open the door wide enough to allow sufficient access for cleaning but avoid pushing the door wide open, especially if the cat might be likely to attempt escape.
- If the cat is friendly and seeks attention, stroke or play with the cat while cleaning.
- If the cat is shy or fearful, allow it to retreat to its hiding place and drape a towel over this to increase the cat's ability to hide and feel safe.
- If cleaning with the cat in the cage is not possible, e.g. if the cat is so fearful that there is a very high chance of escape and/or aggression toward the person cleaning the cage, provide a carrier or something similar as the cat's regular hiding place within its cage, so that the cat can be safely enclosed. The cat can then be allowed to retreat into, or be encouraged into, this so that it can be removed with the cat

inside while cleaning takes place. While the cat is out of the cage and in the carrier, a towel, preferably one from the cage containing the cat's own scent, should be draped over the carrier to provide extra security for the cat.
- Try to maintain the cat's scent profile within its cage. Scent is important, and recognition of its own scent can help to make a place feel familiar and safe to a cat. Care should therefore be taken not to remove this scent during cleaning by employing a 'spot cleaning' method as follows:
  - Only change or wash bedding if it is sufficiently soiled that it could be harmful to the cat's health.
  - It may help to have more than one layer or piece of bedding in the cage so that if one needs to be removed, others containing the cat's scent can remain.
  - Clean areas that have been soiled with urine, faeces, blood or food, but leave other areas untouched, especially where the cat may have deposited scent signals by rubbing. These areas may be recognized as a brown/black slightly oily mark.

Removing a cat from a hospital cage

- Ensure all doors and windows are closed and any potential escape routes are blocked. Also remove litter trays, and food and water dishes from the cage.
- Observe the cat to assess its temperament (see Chapter 3) before approaching and attempting to remove it.
- If the cat appears calm, confident and friendly, approach calmly and slowly from the side, avoiding prolonged direct eye contact.
- Avoid sudden movements, and sharp or loud sounds.
- Speak to the cat in a calm measured voice.
- Offer your hand for the cat to sniff before attempting to stroke the cat.
- If the cat is happy to be stroked and appears calm, only then attempt to pick the cat up and remove it from the cage by supporting it well underneath and lifting the cat out of the cage and securely into your arms.
- If the cat retreats into its hiding box it can often be less stressful for the cat and easier for the handler to remove the box or bed with the cat inside.
- If the cat is unhappy about being picked up and does not retreat into its hiding box, a towel or the cat's bedding can be used to gently wrap the cat before attempting to remove it.
- If the cat is likely to be aggressive it is advisable to wear gauntlets to ensure your safety.
- If the cat is to be placed into a carrier it can be a good idea, if possible, to replace the cat's hiding box with the carrier and then gently encourage the cat into the carrier.

Returning home following hospitalization

Problems can sometimes arise when a cat returns home after a period of hospitalization. For example, the cat may have difficulty in readjusting to life back home, or there could be a disruption in the returning cat's relationship with other cats that live in the same household. The reason for this could be due to any or all of the following:

- The cat may be in pain or discomfort following injury or surgical procedures.
- The cat may still be in a heightened state of stress and arousal following the stay in hospital and the journey home.
- The returning cat will now smell of 'the vets', which could cause a disruption to its relationship with other cats in two possible ways:
    - If the other cat(s) have themselves had bad experiences at the vets, they may associate the scent that is now on the returning cat with potential threat.
    - The returning cat will no longer smell familiar and so might be considered as a stranger and as an intruder to the core territory.
- The relationship between the cats before they were separated may have already been one of resigned tolerance rather than close bonding. Any sense of rivalry and competition for resources that existed previously is likely to increase after a period of separation.

Reducing the possibility of problems following hospitalization

- Ensure that the cat has been prescribed correct and sufficient analgesia and other medications to limit pain and discomfort. Also ensure that the owners know how to, and are capable of, administering the medication (see Appendix 13).
- Try to limit stress during hospitalization and educate the clients on how they may limit stress associated with travelling for the cat (see Appendix 12).
- If the cat's carrier is not needed at the surgery while the cat is hospitalized, it can be better if it is stored at home so that it retains the familiar home scent.
- On arrival home the returning cat may need to be housed in a separate room away from other household pets for a short while (this may also be necessary for medical reasons) and then re-introduced gradually as you would a new cat to the household (see Appendix 5).
- Installing a synthetic feline pheromone diffuser might help the cat to settle. An F3 synthetic facial pheromone diffuser would be most suitable to help a single cat settle back home, whereas the addition of a cat-appeasing pheromone diffuser might help to facilitate reintegration into a multi-cat household (see Pheromonatherapy in Chapter 11b).
- Care should also be taken to limit rivalry and competition for resources between the cats as issues relating to this might not become evident until after they have been separated for a while and are then reunited (see Appendix 3).

International Cat Care has produced some excellent videos on interacting and handling cats in the veterinary clinic, which can be accessed via the following link: www.icatcare.org/cat-handling-videos

References

Cannon, M. and Rodan, I. (2016a) The cat in the veterinary practice. In: Rodan, I. and Heath, S. (eds) *Feline Behavioural Health and Welfare*. Elsevier, St Louis, Missouri, USA, pp. 101–111.

Cannon, M. and Rodan, I. (2016b) The cat in the consultation room. In: Rodan, I. and Heath, S. (eds) *Feline Behavioural Health and Welfare*. Elsevier, St Louis, Missouri, USA, pp. 112–121.

Gouin, J.P. and Kiecolt-Glaser, J.K. (2011) The impact of psychological stress on wound healing: methods and mechanisms. *Immunology and Allergy Clinics of North America* 31, 81–93.

Gourkow, N., LaVoy, A., Dean, G.A. and Phillips, C.J. (2014) Associations of behaviour with secretory immunoglobulin A and cortisol in domestic cats during their first week in an animal shelter. *Applied Animal Behaviour Science* 150, 55–64.

Habacher, G., Gruffydd-Jones, T. and Murray, J. (2010) Use of a web-based questionnaire to explore cat owners' attitudes towards vaccination in cats. *Veterinary Record* 167, 122–127.

Padgett, D.A. and Glaser, R. (2003) How stress influences the immune response. *Trends in Immunology* 24, 444–448.

Sparkes, A., Bond, R., Buffington, T., Caney, S., German, A., Griffin, B., Rodan, I. (2016) Impact of stress and distress on physiology and clinical disease in cats. In: Ellis, S. and Sparkes, A. (eds) *ISFM Guide to Feline Stress and Health: Managing Negative Emotions to Improve Feline Health and Wellbeing*. International Cat Care, Tisbury, Wiltshire, UK, pp. 41–53.

Vogt, A.H., Rodan, I., Brown, M., Brown, S., Buffington, C.T., Forman, M.L., Neilson, J. and Sparkes, A. (2010) AAFP-AAHA feline life stage guidelines. *Journal of the American Animal Hospital Association* 46, 70–85.

Wallinder, E., Hibbert, A., Rudd, S., Finch, N., Parsons, K., Blackwell, E. and Murrell, J. (2012) Are hospitalised cats stressed by observing another cat undergoing routine clinical examination? *ISFM Proceedings* 2012. *Journal of Feline Medicine and Surgery* 14, 650–658.

11b Advice for Veterinary Professionals
Part 2 – Advising Clients: Prevention and Treatment of Feline Behaviour Problems

Veterinary professionals, especially nurses/technicians, can be in an ideal position to advise and educate clients on how feline behaviour issues may be avoided or managed. But first it is important to have sufficient knowledge yourself, gained from a reputable source. Hopefully this book and others listed in the recommended reading list at the end of the book can help, but further study is always advisable. The International Society of Feline Medicine (ISFM) is a reputable organization which is free to join (https://icatcare.org/isfm-membership) and provides a worldwide education and information resource, including free journals, for vets and nurses/technicians. ISFM also runs a distance education certificate course in feline behaviour specifically designed for vets and nurses/technicians (https://icatcare.org/learn/distance-education/behaviour/advanced). Relevant information, advice and education can then be passed on to clients in a number of ways.

Talks

Talks for clients, by members of the practice and/or invited guest speakers, can be a part of practice open days or education evenings and can be targeted at specific groups, e.g. breeders, new cat or kitten owners, or to the general cat-owning public. As well as passing on important information, these sessions can provide an opportunity for clients to meet and get to know the practice staff and each other and can help to promote the practice and increase client 'bonding'.

Nurse/Technician Behavioural Clinics

These provide an opportunity for one-to-one advice to be given to the client and can be most useful when practical 'hands on' procedures need to be demonstrated and practised by the client, such as:

- Training the cat or kitten to become accustomed to veterinary examination (see Appendix 14).
- Teaching the client how to medicate their cat (see Appendix 13).

Handouts

Pre-prepared handouts can supplement advice given during talks and behavioural clinics, and can help the client to remember and comprehend the verbal information they have been given. Handouts can also be invaluable when there is only time for cursory advice to be given, for example during a consultation for other issues or when speaking to the client at the reception desk. It is always preferable, however, that some verbal instruction is also given because it cannot be guaranteed that a client will read the handout.

Preparing a client handout

- Ensure that the information on the handout is correct and from a reputable source. But do not directly copy other people's work unless you have their permission, and then give appropriate credit and indicate if any changes have been made.
- Keep advice simple, straightforward and easy to understand. Try to avoid terminology unless you can provide a simple explanation/definition.
- Break up text by using short paragraphs and/or bullet points.
- Pictures and illustrations can also make the handout look more attractive, interesting and inviting for the client to read.

Behavioural First Aid

This is advice offered to the owner of an animal with a behavioural problem until a full behavioural consultation can be conducted. The aims of behavioural first aid advice are to:

- Offer short-term relief; the advice may not completely resolve the problem or be appropriate for the client long-term, but it may alleviate the worst aspects of the problem until professional help can be obtained.
- Prevent worsening of the problem, for example by advising against anything that the owners may be doing that might exacerbate the behaviour or lead to more serious issues.
- Promote improvement, for example: by providing the owner with some explanation of why the problem may be occurring and dispelling myths and misperceptions; by promoting good practice and advising against bad or ineffective actions; and by helping the client in the right direction towards further help and advice.

Why might behavioural first aid advice be necessary?

Time and resource limitations

Correctly identifying the cause of an unwanted behaviour, devising an appropriate treatment programme, and passing on the necessary information and advice to the owner can be a complex and time-consuming task that usually requires:

- Veterinary examination to rule out or treat possible contributory medical issues.
- Gathering of a full behavioural and medical history.

- Gathering of all relevant information about the cat's environment – a home visit may be required.
- Time, and sufficient knowledge of feline ethology to be able to counsel the client, sensitively, non-prescriptively and accurately.

The average length of a full feline behaviour consultation is around 2–3 hours. First aid advice is often required because when the owner initially asks for help and advice it is highly unlikely that the veterinary surgeon or nurse/technician will have the time and necessary resources available at that point to counsel the client fully.

Client expectations

Clients may not be aware of the necessary complexity of a full behavioural consultation and may expect the behavioural issue to be resolved quickly and easily. They might also expect the vet or nurse/technician to be fully knowledgeable in feline behaviour, especially if they have been advised, albeit correctly, that the veterinary practice is the first place to go to if their pet is exhibiting changes in behaviour but have not been advised why, i.e. that the behavioural changes may be a sign of disease, which needs to be ruled out or confirmed and treated before a full behavioural investigation can begin.

The client's needs

- Owning a pet that is exhibiting undesirable behaviour problems can be very distressing, especially if the owner does not understand the reasons for the behaviour or how it may be managed.
- Pet behaviour problems can cause a breakdown in the pet–owner relationship and have damaging effects on the owner's relationships with family, friends and partners. For example, people may be reluctant to visit if the house smells of cat urine or if the cat is aggressive towards them. Partners may argue as to the best way to deal with the cat's behaviour or even if they are willing to keep the cat.
- Owners may also suffer physical injury from the cat and/or substantial damage to property and subsequent financial costs.
- It is also possible that the owner has lived with the problem for some while and is now becoming increasingly desperate for help. Reasons why owners may not have sought help before can include:
  - Being unaware that help is available. This can be especially the case with cat behaviour problems.
  - Being unaware of who or where to go to for help.
  - Having already tried ineffective advice from friends, family, the internet and/or social media.

The animal's needs

Behaviour problems can be sign of poor behavioural and emotional welfare, plus changes in behaviour can be an indicator that the cat may be unwell or is suffering distress, which can itself be an underlying or trigger factor in a number of physical conditions (see Chapter 6).

Pets with behavioural problems can also be at increased risk of:

- Abandonment.
- Relinquishment to a shelter.
- Euthanasia.
- Physical abuse, often due to owner frustration resulting in ineffectual attempts at punishment or reprimands for the behaviour.

Dos and don'ts of providing first aid behavioural advice

- It is important that the client is made aware that it is only first aid advice being offered and that the advice given may not resolve the issue completely or long term.
- Any advice given must be from a reputable source and based on sound scientific principles.
- Be honest and be aware of limitations, such as knowledge, time and resources.
- Make the client aware of the importance of investigating possible underlying medical conditions. If not able to do so at the time, make an appointment for this to be done.
- Don't aim to diagnose or 'cure' the problem. When the full history of the problem is gathered, the issue may turn out to be more complicated or different than it first appeared.
- Don't guarantee effectiveness of advice or products offered.
- Use handouts (see above regarding guidelines on how to prepare handouts, also see Appendix 15).
- Arrange for the client and their pet to have a full behavioural consultation with a suitably qualified and experienced behaviourist, either 'in-house' or via referral.
- Direct clients to helpful and reputable internet sites (e.g. https://icatcare.org/advice/problem-behaviour; www.catexpert.co.uk/cats/).

Referral or Treat 'In-house'

Sometimes simple advice and increasing owner understanding of feline behaviour can be sufficient; however, feline behaviour cases can vary greatly in severity and complexity and often further behavioural investigation and counselling is required. The options can be to offer this service within the veterinary practice or to refer to an external behaviourist. The essential factors influencing this choice should include:

- The local availability and quality of referral behavioural services (see the section 'Who to Refer to?').
- The ability of the practice to provide a level of service that should be at least equal to available referral services. This will usually require:
  - Appropriate and sufficient experience and qualifications in feline behaviour.
  - The option to preferably make a home visit or access to a suitably equipped space to conduct a behavioural consultation. Any room used for feline behaviour

consultations within the veterinary practice should: be escape-proof so that the cat can be allowed the freedom to explore; have available a white board or similar so that the owner can draw relevant maps of the home environment; contain comfortable seating for the behaviour counsellor, the owner and all family members that attend the consultation; have water, tea and coffee making facilities available; and have no risk of disturbance during a consultation.
- There should be sufficient time available:
  - To conduct a full behavioural consultation, which can take 2–3 hours.
  - To prepare a full written report for the client following the consultation, detailing analysis of the behaviour problem and behaviour modification advice.
  - To provide the client with continued advice and support as required.

The code of conduct of The Royal College of Veterinary Surgeons (RCVS) states that a veterinary surgeon should recognize when a case or treatment is out of their area of competence and be prepared to refer to an individual with the necessary skills and expertise (RCVS, 2017). Although this primarily relates to medical or surgical cases, the same should also apply to behaviour cases. The best option for the client and their pet should always be the primary consideration.

If a decision is made to treat a behaviour case 'in-house' it must be closely monitored and if the behaviour problem increases or is not improved within the expected timescale, or the client wishes to investigate the issues at a level that the practice is unable to offer, then the option to refer, if available, should be offered.

Who to Refer to?

If a feline behaviour case is to be referred it should be to someone who can offer more in terms of relevant education, experience and practical help than can be offered by the veterinary practice. It is important to be aware that if the person the pet is referred to is not also a veterinary surgeon, the referring vet will retain duty of care. Therefore, both the referring veterinary surgeon and the client need to feel assured that the person is suitably qualified, and that any treatment or advice given will be based on the correct identification of the underlying cause of the behaviour and that both the evaluation and treatment will be based on scientific and ethical principles. Identifying a suitable person to refer a feline behavioural case to can, however, be a minefield.

- There is currently no statutory requirement regarding who can treat companion animal behaviour problems. An impressive title or assertions of ability made by an individual are not sufficient evidence that they have the necessary education or experience.
- The qualifications of a potential behaviourist should be investigated to ascertain if they are current, relevant and/or sufficient. Unfortunately, this may not be an easy task as there are now numerous course providers offering qualifications at varying standards in the field of companion animal behaviour.
- Membership or accreditation by an animal behaviour group or organization may provide some indication of education and ability. However, there are several organizations in existence with variable requirements for membership. Competent and reputable behaviour practitioners are more likely to belong to, or be accredited by, an organization that sets strict academic and clinical standards, and ensures that members are insured, adhere to a code of conduct and

regularly update their knowledge by undertaking continuing professional development (CPD).
- The behavioural repertoire and motivations to perform behaviours can vary greatly between species. Some clinical animal behaviourists are equally knowledgeable and competent with the behaviour of more than one species. This does not apply to all, however, and just because an individual has previously demonstrated or is known for having a high level of behavioural knowledge and experience of one species, it should not be assumed that they also have a similar level of knowledge and experience of other species.

The Animal Behaviour and Training Council

Following a report presented by the Companion Animal Welfare Council (CAWC) that outlined the need for regulation, The Animal Behaviour and Training Council (ABTC) was developed as a regulatory body for animal trainers and animal behaviour therapists (Fig. 11.5). Founder members of the council include national and international animal welfare and animal behaviour organizations including: the British Small Animal Veterinary Association (BSAVA); European College of Animal Welfare and Behaviour Medicine (ECAWBM); Royal Society for the Prevention of Cruelty to Animals (RSPCA); The Canine Behaviour and Training Society (TCBTS); The Association for the Study of Animal Behaviour (ASAB); The British Veterinary Behaviour Association (BVBA); and the Association of Pet Behaviour Counsellors (APBC). As well as promoting animal welfare in relation to interactions with people and the education of the animal-owning public, the ABTC maintains national registers for animal training and behaviour practitioner roles, and recognizes individuals and organizations whose members they consider to be suitably qualified to fulfil those roles (www.abtcouncil.org.uk).

Practitioner organizations include The Association of Pet Behaviour Counsellors (www.apbc.org.uk) and The International Association of Animal Behavior Consultants (IAABC; https://iaabc.org).

Behavioural Pharmacology

Psychotropic medication, which can only be prescribed by a veterinary surgeon, can sometimes be useful to aid behaviour therapy by helping to reduce or control emotional responses such as fear, anxiety, arousal, impulsivity and reactivity. But drug therapy in isolation without identifying and correctly addressing the underlying cause of a behaviour is rarely fully effective in the long term.

There are no psychotropic medications licensed for cats in the UK and very few published therapeutic clinical trials (DePorter *et al.*, 2016). Therefore, it is important

Fig. 11.5. The Animal Behaviour and Training Council (ABTC) is a regulatory body representing animal trainers and animal behaviour therapists in the UK.

that any cat prescribed medication for a behaviour issue is closely monitored to gauge improvement and to watch for adverse effects on health. One disadvantage of using medication in cats can be the difficulty in administering it, which can itself be a source of stress for a cat, and should therefore be a major consideration when evaluating the possible benefits or otherwise, and which medication to use.

The most regularly prescribed medications are those that influence behaviour via effects on neurotransmitters, primarily serotonin, dopamine, noradrenaline, acetylcholine, γ-aminobutyric acid (GABA) and glutamate, and their receptors in the central nervous system (see Table 11.1). The precise mechanism of each drug varies, as does the neurotransmitter targeted, but their action is primarily centred on the communication from one nerve cell to another via the neuronal synapse (Fig. 11.6).

Pheromonatherapy

Scent is a prominent part of feline communication involving pheromones, which are chemical substances that convey messages between individuals of the same species (see Chapter 3). Synthetic versions of some of these feline pheromones are available commercially and can be used to help cats cope with stressors or deal with potentially stressful situational events, and/or to assist with behaviour therapy when a behaviour problem already exists.

Synthetic facial pheromones

Five feline facial pheromones (F1–F5) have been identified, two of which, F3 and F4, have been synthetically reproduced. The F3 fraction has been identified as being associated with facial marking of familiar and safe areas. The synthetic analogue of this faction is marketed under the trade name of Feliway Classic and is available as a spray or in a plug-in diffuser that should be located near to the cat's preferred resting place. The spray has an alcohol base that may be unpleasant for the cat, so a few minutes should lapse between applying the spray and allowing the cat access to the area. When using the diffuser, there have been a number of anecdotal reports of cats urine marking on or close to the diffuser when it is first plugged in. This may be due to the scent of the diffuser warming up or the cat perceiving the pheromone scent as a threat before it becomes accustomed to it. To avoid this, it can therefore be advisable to initially plug the diffuser into an electrical socket that is out of the cat's reach and then relocate it after a day or so.

The F3 fraction as a spray or diffuser can be used to help minimize stress for cats at potentially stressful times; for example:

- Moving home: plug in a diffuser in the old home a week or so prior to moving, and then a day or two before taking the cat to the new home, plug a diffuser in at the new house.
- Travelling: spray in the carrier and around the car 10–15 minutes prior to a journey.
- Fireworks or other potentially frightening events: plug in the diffuser a few days before the event is expected.
- Changes to the environment, e.g. decorating, building works: plug in the diffuser a few days before work commences.
- Veterinary/cattery visits: use the spray in the carrier; diffusers can also be installed at the vet practice or cattery.

Table 11.1. Psychopharmacology.

	Description/action	May be used for	Contraindications and precautions
Tricyclic antidepressants (TCAs), e.g. amitriptyline, clomipramine	Block the reuptake of serotonin and noradrenaline from the neuronal synapse and thereby increase the amount of both neurotransmitters available to reach the postsynaptic receptor	Repetitive/compulsive behaviours and where anxiety is an emotional component	Should never be given with monoamine oxidase inhibitors (MAOIs). Avoid using, or use only in very low doses if other serotonin-enhancing medications or supplements are also being given. Use with care in cats that are known to suffer seizures
Selective serotonin reuptake inhibitors (SSRIs), e.g. fluoxetine	Selectively block the reuptake of serotonin from the neuronal synapse	Behaviours where anxiety and impulsivity are indicated	Should never be given with MAOIs. Avoid using, or use only in very low doses, if other serotonin-enhancing medications or supplements are also being given
Azapirones, e.g. buspirone	Serotonin receptor agonists. Bind to and activate the serotonin 1A receptor in the postsynaptic cell	Issues of anxiety, fear, timidity and social tension	Use with caution in cats with renal or hepatic disorders
Benzodiazepines, e.g. alprazolam, clorazepate diazepam	GABA receptor agonists. Facilitate the action of the inhibitory transmitter GABA resulting in decreased neurotransmission throughout the central nervous system (CNS). Behavioural changes due to influences within the hypothalamus and limbic system	Anxiety, fear, hyperarousal. Sedative and anticonvulsant action, with immediate onset of action. May be prescribed for short-term events or in combination with long-term maintenance medications that take a few weeks to reach full therapeutic effect	Use with caution in cats. Cases of fatal hepatic necrosis have been reported. Overdose can cause severe CNS depression in cats with hepatic disease
Gabapentin	Inhibits the release of excitatory neurotransmitters, particularly substance P, glutamate and noradrenaline. Has anticonvulsant and analgesic effects	Useful for hypersensitivity, anxiety or aggression, especially if pain is also a contributory factor	May cause mild sedation or ataxia but generally well tolerated. Xylitol (a sweetener) in gabapentin preparations for human use may cause hypoglycaemia or hepatotoxicity and therefore should be avoided

Continued

Table 11.1. Continued.

	Description/action	May be used for	Contraindications and precautions
Monoamine oxidase B inhibitors (MAOBIs), Selegiline	Inhibits the actions of the enzyme monoamine oxidase B which catabolizes (breaks down) catecholamines, including dopamine, adrenaline and serotonin in the CNS	May be prescribed in cases of cognitive dysfunction in older cats (Gunn-Moore, 2011)	Should never be used in combination with TCAs, SSRIs or other medications including tramadol, metronidazole, prednisolone or trimethoprim
Megestrol acetate	A powerful progestogen (hormone) that can delay or prevent oestrus in entire female cats	Has been used to treat urine spraying; however, the damaging side-effects outweigh the potential benefits	Common serious side-effects of weight-gain, mammary hyperplasia/neoplasia, blood dyscrasia (abnormal blood pathology) and increased risk of diabetes

(Crowell-Davies and Landsberg, 2009; Gunn-Moore, 2011; DePorter et al., 2016.)

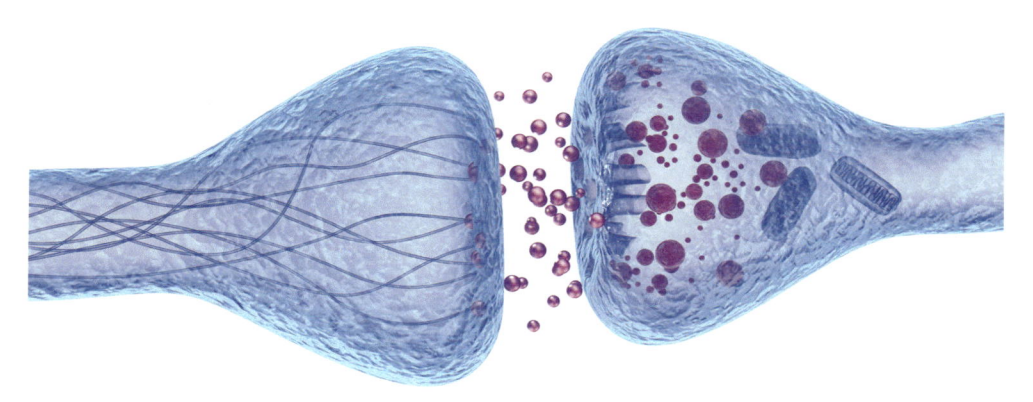

Fig. 11.6. Messages are conveyed from one neuron to another in chemical form via a small gap known as the synapse. These chemicals, known as neurotransmitters, are released from the presynaptic membrane on the axon terminal but not all cross the synaptic cleft because some are resorbed back into the axon terminal (known as reuptake). The neurotransmitter molecules that do cross the synaptic cleft then bind with their specific receptors in the postsynaptic membrane.

The F3 fraction can also be used, alongside specific behavioural advice, to help with behaviour problems such as:

- Any changes in behaviour related to stress.
- Urine marking: use the spray on the areas where the cat is urine marking (see Appendix 15).
- Furniture scratching: excessive indoor scratching may be due to stress; if so the scent from a diffuser may help to reduce this. Also applying the spray onto surfaces where you do not want the cat to scratch can help to relocate the scratching to other areas.

The F4 fraction has been identified as the scent exchanged during allorubbing and allomarking (rubbing on the same inanimate object) by social group members. The synthetic version, marketed under the name Felifriend, is designed to facilitate handling by rubbing the product into the handler's palms and wrists, and to aid introductions or reintroductions to people or other animals, including other cats. However, there can be a problem in using this product, due to the fact that feline social interactions are not reliant on scent alone. It may, therefore, cause frustration and a sense of conflict if the other signals perceived by the cat do not match the message supplied by the pheromone.

Cat-appeasing pheromone (CAP)

This is a pheromone secreted from the intermammary sulcus of a nursing female that appears to appease and aid cohesion between individuals in a litter of kittens and increase bonding to the mother cat. The pheromone is also reported to have a similar action in adult cats (Cozzi *et al.*, 2010). A synthetic version has been developed and is commercially available as a plug-in diffuser. Sold as Feliway Friends in the UK and Feliway Multi-Cat in the USA this may be used to aid introductions and help to reduce conflict between cats living in the same household. Diffusers should be located in the house in an area shared by the cats, preferably a shared resting area.

Feline interdigital semiochemical (FIS)

Scratching that leaves marks on surfaces is also a form of feline communication. As well as a visual scratch, an olfactory marker is also deposited from the interdigital glands between the toes and from glands in the large foot pad, known as the plantar pad (see Chapter 3). This chemical scent has also been artificially reproduced and made commercially available under the name of Feliscratch. Applying this to a suitable scratch post or pad may help to redirect cats away from items that the owner does not want the cat to scratch and towards a more acceptable surface. The product also contains a blue dye that can help to mimic the visual marking, plus catnip that can also act as an attractant and an encouragement for the cat to scratch on the surface to which it is applied.

Complementary and Alternative Medicine (CAM)

This covers the very wide range of treatments and preparations not generally considered to be a part of conventional medicine, including supplements and herbal products (see Table 11.2) and many other treatment methods and therapies, a few of which are described below. Complementary medicine or therapy refers to treatment that is used alongside conventional health care. Treatments that are used in place of mainstream practice are termed alternative.

Efficacy, quality and safety can vary tremendously, and it should never be assumed that a product or method is safe just because it is presented as 'natural'. For pharmacological drugs to be licensed, a substantial body of proof is required that they are safe and effective but there is very little similar requirement for CAM therapies or products.

Homeopathy

There are three main principles of homeopathy, proposed by Samuel Hahnemann in 1796. The first is that 'like cures like', in other words a substance that causes signs or symptoms in a healthy individual that are similar to the signs or symptoms of a particular disease or disorder will cure or ease the symptoms of an individual afflicted with that disorder. For example, coffee bean remedy is said to cure insomnia and onion remedy to cure the symptoms of a cold.

The second principle applies to the preparation of the remedies, which involves serial dilution, often until there is no detectable amount of the original substance, but it is claimed that the water retains a memory and the higher the dilution the stronger the potency of the remedy.

The third principle also applies to the preparation and is known as 'succussion', a specific form of vigorous shaking or tapping at each stage of dilution, which is also believed to increase the remedy's healing power (Lees *et al.*, 2017a).

Bach flower remedies

These are infusions of plants and flowers preserved in brandy and diluted by spring water. As with homeopathy, the concentration of plant extract in the final preparation

Table 11.2. Examples of nutraceuticals* and herbal supplements that may aid behaviour therapy. (Adapted from DePorter et al., 2016.)

Supplement	How does it work	Indications	Precautions
Alpha Casozepine A nutraceutical derived from casein, a protein in cow's milk	Reported to have anxiolytic effects similar to benzodiazepines	Fear, anxiety, stress	None reported
Tryptophan An essential amino acid (one of the building blocks of proteins) and the dietary precursor of serotonin	May increase available levels of serotonin or enhance its transmission	Anxiety, stress	May increase the risk of serotonin syndrome[‡] if used concurrently with drugs that increase serotonin levels
Theanine An amino acid found in green tea that has a chemical structure similar to glutamate	May block the effects of glutamate and increase GABA. An advantage is that it is highly palatable to cats	Fear, anxiety, stress	None reported
Docosahexanoic acid (DHA) and Eicosapentaenoic acid (EPA) Polyunsaturated fatty acids	Essential for early brain and retinal development, and to help maintain brain function in adults	For healthy brain and retinal development in kittens and to improve cognitive function in senior cats	Avoid high doses, which may result in gastrointestinal effects, weight gain, altered immune function and poor wound healing (Lennox, 1993)
Valerian Herb *Valeriana officinalis*	Usually prepared with other herbs in either a tablet or liquid form for oral administration or in a spray or diffuser for environmental use. Can have sedative properties	Anxiety, fear, stress, excitability, timidity, and car sickness	
Catnip Herb *Nepeta cataria*	The active ingredient is the essential oil nepetalactone which affects the CNS via the olfactory bulb. Does not affect all cats (only around 50-70%) and effects can be variable with affected cats showing a wide range of behaviours	Encouragement to play and aid enrichment. May be used to encourage scratching on a suitable surface	May cause increased arousal resulting in short term aggression. Catnip intoxication has been reported

*A nutraceutical is a food or food derivative that is said to have medicinal or health-giving properties.
[‡]Serotonin syndrome is a term used to describe a potentially fatal condition that can arise if serotonin levels or availability are excessively increased.

is undetectable, but it is also claimed that the water retains a memory of the substance. Flower remedies were originally created by Dr Edward Bach (1886–1936) who believed that illness is caused by negative states of mind, which makes them popular in the treatment of animal behaviour issues. The most popular of the remedies, also marketed for use in pets, is 'Rescue Remedy', a combination of five flower remedies designed to reduce anxiety and stress in potentially stressful situations. There is no evidence that these remedies are associated with any therapeutic effects and a randomized, double-blind, placebo-controlled trial of Rescue Remedy in humans demonstrated no evidence of the efficacy of Rescue Remedy over placebo effect (Armstrong and Ernst, 2001; Lindley, 2009).

Zoopharmacognosy

This is the practice of allowing animals to self-medicate by 'selecting' from a range of mainly plant-based substances. If the animal demonstrates a so-called 'positive reaction', which encompasses a wide range of non-specific behaviours, such as sniffing, blinking, stillness or backing away from the substance, the animal is then allowed or encouraged to either ingest the substance, or it is applied to the animal topically, in which case essential oils may be used.

The practice is based on observations that occasionally wild animals will eat plants or other substances that prove to be beneficial for their health, e.g. the ingestion of plants with anthelmintic (antiparasitic) properties, and clay containing kaolin-like substances by primates (Mahaney *et al.*, 1996). This is behaviour that is most likely to have developed over generations via trial and error and observational learning of others. However, as well as ingesting potentially beneficial plants and minerals, wild animals will also eat plants or other substances that cause severe toxicity (Robles *et al.*, 1994), something that is also far from uncommon among companion animals. Care should also be taken with the use of essential oils around cats owing to potential toxicity.

Acupuncture

A treatment derived from traditional Chinese medicine, acupuncture involves the insertion of fine needles into specific parts of the body. Research has shown that the practice may have therapeutic effects via stimulation of the nervous system. It may be indirectly helpful in the treatment of behaviour problems by helping to alleviate contributory physical issues such as chronic pain, although it has also been shown that serotonin, noradrenaline, endogenous opiates and possibly oxytocin are released following acupuncture stimulation (Lindley, 2009).

Aromatherapy

In humans, aromatherapy is used to influence mood and produce a sense of well-being, theoretically by working via the hypothalamus and limbic system, which governs emotional response (Lindley, 2009).

The use of lavender oil has been shown to calm travel-induced excitement in dogs but it is unclear if this is due to a true aromatherapy effect or the distraction of a strong scent on an animal with a highly acute sense of smell (Wells, 2006). A study assessing the effects of a variety of scents, including lavender, presented to cats in a rescue shelter found that the cats demonstrated far more interest in the scent of catnip and, to a lesser extent, the scent of prey (rabbit) than in the lavender scent (Ellis and Wells, 2010).

TTouch

Using principles of the Feldenkrais method of exercise therapy for humans that is said to improve both physical movement and mental well-being, TTouch was developed in 1978 by Linda Tellington-Jones as a specific form of human to animal touch and exercise for animals that can have an influence on the central nervous system. Massage in humans has been shown to decrease levels of cortisol and increase levels of serotonin and dopamine (Field *et al.*, 2005), and similar effects are also likely to occur in animals, which could have a beneficial effect on the negative emotions associated with stress (Cascade, 2012). Whether TTouch is superior to massage techniques is debatable; however, practitioners are trained to observe the animal's response and aim to ensure that the therapy is performed in a manner that is pleasurable and of most benefit to the animal.

The efficacy and safety of CAM

Acupuncture and TTouch appear to have some scientifically plausible mechanism of action, but with others this is debatable. Even so, CAM therapies are increasingly popular, and practitioners and enthusiasts frequently claim that they are effective. Their use in veterinary medicine, and in the treatment of animal behaviour is controversial, however, because apparently successful treatment could also be attributable to other factors (see The Placebo Effect). In the case of homeopathy, successful treatment might be linked to consultation with a true homeopathic practitioner who is more likely to treat the patient holistically, and who may therefore also give beneficial advice regarding diet and husbandry.

Followers of these therapies also claim that 'at least they do no harm', which may be true as far as some of the treatments are concerned, especially the very high dilutions used in homeopathy and Bach's Flower Remedies. However, not all homeopathic remedies are administered at high dilutions and, if given in a more concentrated form, the source material could be potentially hazardous (Lees *et al.*, 2017b). With zoopharmacognosy there is very real potential for harm. Just because an animal chooses to eat or come into close contact with a substance does not guarantee that it is not harmful to them. This is especially the case with cats because there are a great many plants and their derivatives that are toxic to cats (see: https://icatcare.org/advice/poisonous-plants). Plus, because they are specialist carnivores, cats lack the liver enzymes that breakdown harmful chemicals, so once a cat has ingested a toxic substance it often has a poor chance of recovery. Even if the substance is not fed to the cat and just applied to the skin, the cat's normal grooming behaviour will usually result in the substance being ingested.

Another way in which these therapies might be harmful is if they are used as an alternative rather than alongside conventional treatment, which may delay or completely deny the animal access to necessary conventional veterinary or behavioural treatment.

The Placebo Effect

The placebo effect is when an improvement in health, behaviour, or a reduction of signs or symptoms occurs that cannot be attributed to the treatment that has been undertaken. During drug trials, a group of individuals is normally given a fake medication that appears similar in every other way to the 'real' drug except that it contains no therapeutic properties. Despite this, a significant number of the individuals receiving the fake medication will report an improvement in their symptoms. When the placebo effect occurs in humans, expectation of improvement can be a simple explanation but this clearly cannot occur in animals. However, some authors and studies report that placebo effect, or something very similar, does occur in animals (McMillan, 1999). Simple coincidence can be a factor, i.e. the animal's health or behaviour would have improved anyway, plus, in the field of animal behaviour especially, there is also the high possibility of 'indirect or vicarious efficacy', whereby an undesirable behaviour by the animal is triggered or closely linked to the owner's behaviour (Lees *et al.*, 2017a). In other words, when the animal is 'on medication' the owner no longer feels the need to perform a behaviour that may have triggered or contributed to the undesirable behaviour in the animal, so the animal's behaviour improves. For example, the owner may feel the need to shout at the cat whenever he suspects that the cat is about to spray urine on the furniture, but it is being shouted at by the owner that triggers the urine marking. Once the cat is on medication to 'cure the urine marking' the owner no longer feels the need to shout at the cat so the behaviour problem is resolved.

The placebo effect is not yet fully understood, and there are several theories as to why an animal's health or behaviour might improve without it being as a direct result of medication, conventional or otherwise.

References

Armstrong, N.C. and Ernst, E. (2001) A randomized, double-blind, placebo-controlled trial of a Bach Flower Remedy. *Complementary Therapies in Nursing and Midwifery* 7, 215–221.

Cascade, K. (2012) The Sensory Side of TTouch. Available at: http://www.ttouchtteam.org.uk/ArtKathyCascadeSensory.shtml (accessed 31 January 2018).

Cozzi, A., Monneret, P., Lafont-Lecuelle, C., Bougrat, L., Gaultier, E. and Pageat, P. (2010) The maternal cat appeasing pheromone: exploratory study of the effects on aggression and affiliative interactions in cats. *Journal of Veterinary Behavior: Clinical Applications and Research* 5, 37–38.

Crowell-Davies, S.L. and Landsberg, G.M. (2009) Pharmacology and pheromone therapy. In: Horwitz, D.F. and Mills, D. (eds), 2nd edn. *BSAVA Manual of Canine and Feline Behavioural Medicine.* BSAVA, Gloucester, UK, pp. 245–258.

DePorter, T., Landsberg, G.M. and Horwitz, D. (2016) Tools of the trade: Psychopharmacology and nutrition. In: Rodan, I. and Heath, S. (eds) *Feline Behavioural Health and Welfare*. Saunders, Elsevier, Philadelphia, Pennsylvania, USA, pp. 245–267.

Ellis, S.L. and Wells, D.L. (2010) The influence of olfactory stimulation on the behaviour of cats housed in a rescue shelter. *Applied Animal Science* 123, 56–62.

Field, T., Hernanadez-Reif, M., Diego, M., Schanberg, S. and Kuhn, C. (2005) Cortisol decreases and serotonin and dopamine increase following massage therapy. *International Journal of Neuroscience* 115, 1397–1413.

Gunn-Moore, D. (2011) Cognitive dysfunction in cats: clinical assessment and management. *Topics in Companion Animal Medicine* 26, 17–24.

Lees, P., Chambers, D., Pelligrand, L., Toutain, P.-L., Whiting, M. and Whitehead, M.L. (2017a) Comparison of veterinary drugs and veterinary homeopathy: part 1. *Veterinary Record* 181(7), 170–176. DOI: 10.1136/vr.104278.

Lees, P., Chambers, D., Pelligrand, L., Toutain, P.-L., Whiting, M. and Whiteghead, M.L. (2017b) Comparison of veterinary drugs and veterinary homeopathy: part 2. *Veterinary Record* 181(8), 198–207. DOI: 10.1136/vr.104279.

Lennox, C.E. and Bauer, J.E. (2013) Potential adverse effects of omega-3 fatty acids in dogs and cats. *Journal of Veterinary Internal Medicine* 27(2), 217–222.

Lindley, S. (2009) Complementary therapies therapy. In: Horwitz, D.F. and Mills, D. (eds) *BSAVA Manual of Canine and Feline Behavioural Medicine*, 2nd edn. BSAVA, Gloucester, UK, pp. 259–269.

Mahaney, W.C., Hancock, R.G.V., Aufreiter, S. and Huffman, M.A. (1996) Geochemistry and clay mineralogy of termite mound soil and the role of geophagy in chimpanzees of the mahale mountains, Tanzania. *Primates* 37, 121–134.

McMillan, F.D. (1999) The placebo effect in animals. *Journal of the American Veterinary Medical Association* 1, 992–999.

RCVS (2017) Code of Professional Conduct for Veterinary Surgeons. Available at: www.rcvs.org.uk/advice-and-guidance/code-of-professional-conduct-for-veterinary-surgeons (accessed 31 January 2018).

Robles, M., Aregullin, M., West, J. and Rodriguez, E. (1994) Recent studies on the zoopharmacognosy, pharmacology and neurotoxicology of sesquiterpene lactones. *Planta Medica* 61, 199–203.

Wells, D. (2006) Aromatherapy for travel-induced excitement in dogs. *Journal of the American Veterinary Medical Association* 229, 964–967.

12 Advice for Other Cat Carers

Owners, breeders, and veterinary staff are not the only people involved with the care of cats. Cat carers also include cattery owners/workers, shelter staff, cat sitters and anyone involved with the recent and increasing trend of cat cafés. Admittedly all these roles are worthy of their own 'advice' chapters but unfortunately there is not enough space to allow for this. Therefore, I have combined advice for all other cat carers into one chapter.

General Advice for All

Whatever role you are already involved in or considering becoming involved in, it is advisable to gain as much knowledge as you can about feline behaviour and the behavioural and welfare needs of cats. This book will hopefully help to some extent, but further study is always advisable, especially in the areas of feline behaviour that are most relevant to you and your cat carer role. There is a lot of information that can be gathered from books and the internet, but the standard and accuracy of such information can vary. Therefore, it is essential that you get the information you need from a reputable source (see the list of useful websites and the recommended reading list at the back of this book).

Basic behavioural and welfare needs

There are some basic factors of which all cat carers must be aware. All cats need:

- Space: a feral cat can have a natural range of around 2 km^2 (Gourkow *et al.*, 2014). Although most cats can cope with much less, the more severe the confinement the greater the potential damage to the cat's overall welfare.
- The opportunity to engage in natural behaviours.
- To feel safe. All cat carers should be aware of what are real, potential or perceived threats for cats and how to avoid or reduce them, or provide the cat with sanctuary from such threats.
- Predictability and control. Unpredictable care and husbandry procedures can be a major source of stress for cats. It is also important that they feel they have some level of control over their environment and their interactions with others.

Advice for Shelters and Catteries

Location of cat housing

- In a quiet area. Well away from sounds that may be perceived as threatening or disturbing; for example, barking dogs, machinery, loud sounds of human origin – traffic noise, loud music, children playing, etc.

- Away from bright, flashing or harsh artificial lighting.
- Avoiding extremes of temperature (local licensing regulations must be followed). Pens should be well ventilated and kept warm in cold weather and cool in hot weather, so that a fairly constant and comfortable ambient temperature is achieved. Ideally, individual outdoor space should also be available that allows the cat access to sun, shade, and fresh air.

Construction, design and furniture

Following local licensing regulations as a minimum guideline, sufficient space within each pen must be provided (local licensing regulations must be followed). Vertical space, i.e. access to shelves or surfaces at varying levels, is particularly important. Being in an elevated position can help a cat to feel more in control and less threatened. However, it is also important that access to elevated areas is not challenging or difficult, especially for cats that have reduced mobility owing to old age, injury or disease. Ramps or steps should therefore not be too steep and should have non-slip surfaces.

Cats housed in individual units should not be housed where they are in constant sight of other cats. This may be prevented by erecting a solid barrier over cage wire or by applying frosting to glass through which other cats are likely to be seen.

Provision must be made for the cat to be able to access a sectioned off 'private' area. This area needs to be large enough to contain a resting area, food, water and a litter tray, all of which must be positioned well apart from each other. This can often be best achieved by allowing the cat access to different levels within the sectioned-off area.

Hiding places should be provided in addition to the 'private' area (Fig. 12.1). Boxes, solid-sided carriers, igloo beds, covered litter trays converted into beds or even a towel draped over a fitted shelf can be used for this purpose. If a cat is aggressive or appears highly fearful and stressed, increasing the size of the pen and options for hiding often helps.

Fig. 12.1. Hiding places must be provided.

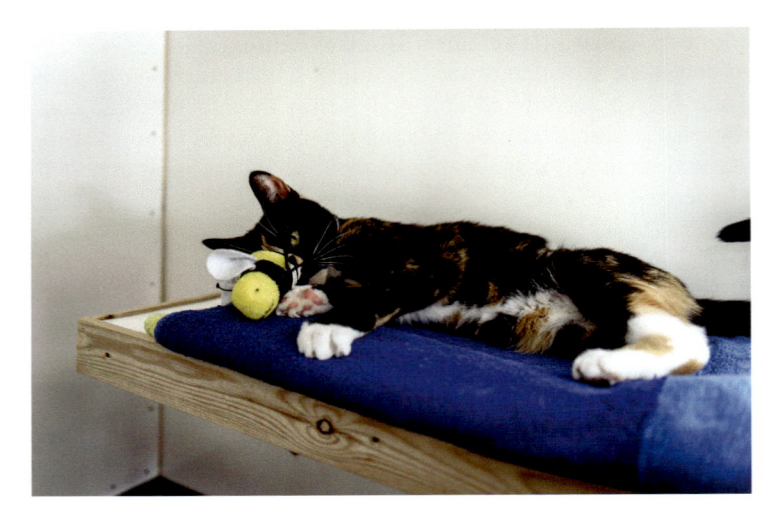

Fig. 12.2. Bedding should be provided in hiding places and on elevated shelves. Being in an elevated position can help a cat to feel more in control and more at ease.

The cat should have access to soft, warm and comfortable bedding and be given options for different places to rest. Bedding should be provided on elevated shelves (Fig. 12.2), as well as in hiding boxes. Unwashed bedding supplied by the owners that has been used previously by the cat can help it settle into a new environment as this will contain familiar and comforting scents.

Provision must also be made for mental and physical stimulation and to allow the cat to engage in natural behaviours by including the following:

- Scratch posts, pads, tree branches or similar, to allow scratching and climbing.
- A view, direct or via a window, of an area containing trees, bushes, etc., where birds or other wildlife may be seen.
- Puzzle feeders to increase enrichment and stimulation (e.g. www.foodpuzzles forcats.com). Cats should be monitored, however, to ensure that using food puzzles is not increasing or causing frustration, or resulting in the cats being unable to access all their food requirements.

Design considerations

- A popular design for shelter or cattery pens can be to have two separate areas with access between the two via a cat flap. Although this can be a good idea for cats that are accustomed to and comfortable using a cat flap, others may not have used a cat flap previously or they may be wary of using one in an unfamiliar environment.
- Glass-fronted 'pod' designs are currently popular for sleeping areas. Although glass is good for disease control and allows visitors and prospective owners to view the cats, it denies the cat much-needed privacy. Plus, it prevents the cat from gaining essential olfactory (scent) information about people and anything else on the other side of the glass. Because a cat's sense of smell is one of its foremost ways of gaining information, this can be potentially stressful.

Housing two or more cats together

If two or more cats come in from the same household it can seem sensible to house them together in the same pen, but just because they live together does not automatically mean that they get along well, and confining them together could further disrupt their relationship. It is therefore advisable to question the owner carefully to get a good picture of the cats' relationship before placing them in the same pen (see Appendix 16). If there is any doubt, house them separately.

One way to test their relationship can be to keep one in its pen while allowing the other access to a secured area just outside the pen, and then observe their behaviour. If they appear to want to be together and shows signs of friendly greeting behaviour (see Appendix 16), then it may be safe to place them together. However, they must then be observed closely and if signs of conflict are seen it may be better to separate them.

Follow the same rules for housing as for communal housing (see below).

Communal housing

In some shelter environments cats are penned together in communal areas. Individual housing is always preferable because of the enormous stressors placed on cats that are forced into close confinement with other cats, especially in shelter environments where most of the cats would be previously unknown to each other. In addition, with constant changes, i.e. cats being found homes and new cats being introduced, there is rarely the opportunity for a stable group to develop. Behavioural inhibition to avoid the very real risk of physical conflict also means that cats in this type of environment may appear, to the untrained eye, to be happy with their situation when they are in reality severely stressed (see Appendix 18). Health issues are also greater for cats housed communally because diseases spread quicker and can be far more difficult to control.

Some shelters, however, especially those in impoverished communities, have no other option than to house at least some of their cats in communal enclosures owing to insufficient space or a lack of finances available to construct individual units. If some good-standard individual housing is available, then it is advisable to carefully select which cats are more likely to adapt to being housed with others and use the individual housing for those that are less likely to cope in such a situation. Communal housing should never be considered as an option for boarding establishments.

Cats that might be more able to cope with being in a communal enclosure are more likely to be:

- Young cats, preferably under a year old.
- Confident and well socialized with other cats.
- Cats that have a history of living amiably with other cats.
- Cats that show signs of friendly behaviour towards other cats.

Cats considered for housing in a communal enclosure must also be healthy, neutered and fully vaccinated. It is also essential to design the communal enclosure so that conflict and stress can be minimized.

Minimizing stress and conflict in a communal enclosure

- The area should be large enough to accommodate not only all the cats and necessary 'furniture', but also to allow space for workers and volunteers to care for and interact with the cats.
- All the cats should be able to access resources without needing to disrupt or get too close to other cats that may be eating or resting. Ideally there should be more resting and feeding areas than the number of cats in the enclosure.
- Food, water, litter trays and resting areas should all be located away from each other.
- A separate food dish for each cat plus extra dishes, or preferably puzzle feeders, should be provided. Do not feed multiple cats from a single dish.
- Ensure that food dishes are positioned so that each cat can see the whole room while eating.
- Competition and conflict over food can be a potential problem in communal housing; ample feeding resources and feeding little and often, rather than 1–2 times a day, can often help to reduce this.
- Water dishes should also be positioned away from food and litter trays. If there are several cats in the enclosure, it may not be necessary to supply a water dish for each cat but ensure that they have ample choice.
- Litter trays are best positioned in the corners of the enclosure so that the cats are partially enclosed, but avoid using covered trays because this can provide an opportunity for one cat to ambush another as it is exiting the covered tray, plus it is not so obvious when covered trays require cleaning.
- In an enclosure with several cats it may not be possible to supply the ideal number of one litter tray per cat plus one extra; therefore ensure that trays are large and cleaned frequently.
- Provide several, at least one for each cat, 'single' shelves (just large enough for one cat) on varying levels, to allow the cats to rest away from the others. Resting areas should be at least 3 feet (1 m) apart.
- Provide a selection of hiding places. These should all have two holes that act as entrance and exit ports. If there is only one hole the cat that is inside may become trapped by other cats or be ambushed as he attempts to leave.
- Provide ample scratching posts, pads and opportunities to climb.
- Try to provide an 'interesting' view outside the pen, ideally containing trees and/or bushes where birds or other wildlife may be seen.
- Closely monitor all cats in the communal area and remove any that show signs of: distress, fear, withdrawal or avoidance of the others, aggression, blocking or chasing.

Introducing a new cat to a communal enclosure

- Place the cat in a large cage within the enclosure. The cage needs to be large enough to contain all the cat's necessary resources – food, water, litter tray and bedding with space in between them. The provision of hiding areas and a solid structure that the cat can sit on top of are also essential. Part of the cage should be covered with a large towel or similar to provide an area that is blocked from the

sight of the other cats in the enclosure. The cat's resources, i.e. food, water, litter tray and bed, should be placed within this blocked off area.
- Keep the new cat in the cage for approximately 24 hours, observing the interactions between the caged cat and the others in the enclosure.
- If predominantly friendly and affiliative behaviours are observed it should be safe to allow the cat to join the others in the communal enclosure.
- Remove the cat immediately if it shows any signs of fear or aggression, or if negative emotional responses are demonstrated by any of the cats already in the enclosure.

Human contact

Cats in a shelter or cattery environment will require varying levels of human contact and interaction. Confident and well-socialized cats are likely to require human contact in the form of affection and play (see Appendix 2), whereas less well-socialized or timid cats may benefit from having the opportunity to become gradually accustomed to a non-invasive and non-threatening human presence.

- All cats, regardless of history, must initially be given sufficient time to settle into their new environment before any attempt is made to interact with them (see housing a new admission).
- The cat must be allowed to initiate contact and always be allowed to move away and hide from the person whenever they want.
- With timid or poorly socialized cats it is better that the person attempts no form of approach or contact until the cat is ready to approach them. It can often be better to simply sit in or near the pen while reading a book and occasionally speaking softly, without attempting to look towards or make any movement towards the cat.
- If the cat chooses to approach, initial interaction should take place as described in Appendix 17.
- Occasional individual access to a larger area that allows greater opportunity for play and interaction with people can be of great benefit for friendly, well-socialized cats and especially for cats showing signs of frustration such as excessive vocalization, pawing through the cage bars at people passing by or destructive behaviour (ripping up bedding, or tipping up food, water and/or litter trays).

Husbandry and general care procedures

Admitting new cats to the shelter or cattery

Stress is likely to be greater for a cat if it is has had a long or bad journey and if subjected to a long wait between arrival and being caged. There may be little that can be done about the journey to the centre, although cattery owners may be able to advise regular clients on how they might be able to reduce stress for their cats associated with travelling (see Appendix 12). However, adapting the admissions process to minimize stress should be possible.

- Make appointments for arrivals or at least agree/arrange an approximate time of arrival so that a suitable pen can be made ready.

- If the centre houses both dogs and cats, aim to have a separate reception/waiting area for cats, or separate times when dogs and cats are admitted.
- Provide shelves in the reception/waiting area that cat carriers can be placed on. Cats can feel more at threat when at ground level, especially whilst confined in a cat carrier.
- Provide towels or similar that can be used to cover cat carriers and help the cats inside to feel more secure.
- Avoid leaning over the cat in the carrier and/or prolonged peering into the carrier at the cat.
- If paperwork needs to be completed, it might be better to house the cat first and then complete the necessary forms, etc.

Housing a new admission

- New admissions should be housed individually, unless they are closely bonded cats from the same household (see the section on housing two or more cats together).
- Leave the cat with minimal disturbance for the first 2–3 days because it can take this long for the cat to settle and become accustomed to its new surroundings. More time might be needed for timid, nervous or less well-socialized cats.
- As well as providing hiding places within the pen, covering the front may also help the cat to relax and settle.
- Do not attempt to handle the cat, unless absolutely necessary, for the first 2–3 days.
- If veterinary treatment or any other intrusive handing is required, this should be performed by a separate handler to the one who regularly handles or cares for the cat, and should not be performed within the sight of other cats or other animals.
- Avoid 'scruffing' or any form of harsh or rough handling. Employ low-stress handling techniques (see Chapter 11a and Appendix 17).

Consistency

For nervous and less well-socialized cats especially, the familiarity of the pen may be the closest thing that the cat has to a place of safety in what it may regard as an unfamiliar and hostile environment. To maintain this, it is best that the cat remains in the same pen during its stay.

Feeding routine

Unpredictability can be a major stressor for cats, so try to keep to a regular daily routine.

Anticipation and delayed reward can be a major cause of frustration. To avoid this, try to limit the time between when food is expected and delivered.

- Pre-prepare all the food in a separate area, preferably away from where the cats are housed and behind closed doors, to limit their awareness that food is being prepared. This can help to reduce the time between the first and the last cat being fed.
- Involve as many members of staff/volunteers as possible so that more cats can be fed at the same time.

Cleaning

Recognition of its own scent can help to make a place feel familiar and safe to a cat. Care should therefore be taken not to remove this scent during cleaning. Thorough cleaning should only take place if the cat is to be housed elsewhere or has left the centre.

- Remove soiled patches in litter trays 2–3 times daily. Clean the tray completely 2–3 times a week.
- Only change or wash bedding if it is sufficiently soiled that it could be harmful to the cat's health, e.g. if it is soiled with faeces, blood, vomit or spoiled food.
- Ensure that there is plenty of bedding supplied in the pen so that if some needs to be removed, there is still some left that contains the cat's scent.
- Do not clean areas where the cat has deposited scent by rubbing. These areas may be recognized as a brown/black slightly oily mark.

The reader is directed to Gourkow (2016a,b) as a source of information on advising catteries and shelters, and for this section.

Foster Care

Short-term care in a home environment can be an invaluable alternative to shelter housing or cattery accommodation, especially for:

- Cats that are unable to cope with being in a shelter or cattery, or are difficult to handle or assess when housed in a pen.
- Kittens that require early socialization and habituation in a domestic environment, which they are much less likely to receive in a shelter environment (see Chapters 5 and 8).
- Cats that require more intensive individual care than it is possible to provide in a cattery or shelter, for example: cats with chronic health problems; elderly cats; pregnant females; or cats with behavioural issues that can only be fully assessed and dealt with in a domestic setting.
- Cats whose owners are unable to care for their pets due to illness (see for example www.cinnamon.org.uk) or who are victims of domestic abuse (see for example https://www.rspca.org.uk/whatwedo/care/petretreat; www.pawsforkids.org.uk).

Compiled with reference to Halls, 2017.

In most cases the fostering organization will pay for food and other necessities such as cat litter and veterinary care. Therefore, fostering can also provide an opportunity for people who are unable to keep a pet long term, or those who have financial restrictions, to enjoy the benefits of pet ownership.

Ideally cat fosterers should:

- Have no other cats or dogs, or young children.
    - Other pets and young children, even if friendly and well behaved, may be perceived as threatening by a cat that is new to the home and may already be in a distressed state.
    - The exception can be if fostering a kitten that requires socialization with other animals or people, in which case it is imperative that the fosterers' animal(s) and/or children are calm, friendly, well-behaved and familiar with kittens.

- Have a home with sufficient space. A spare room that can be used as a safe area for the cat, and during initial integration into the household, will often be required.
- Already have experience of cat ownership, and have some knowledge of feline behavioural and welfare needs or be willing and able to learn.
- Be calm and patient. It may take time to gain the cat's confidence, especially if it is not well socialized or has come from a situation where it might have been subjected to abuse.
- Be prepared for potential issues such as house soiling or parasites being introduced into the house.
- Be flexible as to each individual cat's needs and have the necessary time available to devote to the cat and its care.
- Be willing to receive training and/or mentoring.
- Be willing and able to keep in regular contact with the fostering organization.
- Be able to transport the cat as and when necessary.
- Be willing to 'let go' at the end of the term of fostering.

Compiled with reference to Halls, 2017.

Adoption

The ultimate aim of most rescue organizations is to find suitable and caring homes for the animals in their care.

- It is very important to allow cats in rescue centres to hide but this can mean that the cat may not always be 'on show' for prospective owners. Information about the cat, including a photograph, should therefore be provided on the outside of the pen. It is important, however, that this is not located in a position where people reading the information could be misperceived by the cat as staring into the pen in a threatening manner.
- Keep in mind that the standards of care and husbandry seen in the centre may be replicated by the new owners. For example, very small amounts of cat litter may be used in the centre to minimize costs but if the new owners continue to use such minimal amounts of cat litter this could potentially result in inappropriate house soiling.
- Owners should be questioned as to why they want a cat and be advised sensitively if a rescue cat might not be the best pet for them (see Chapter 9).

Home inspections

A home visit to check the suitability of the new home is advisable, and home visitors should have sufficient knowledge of feline health and behavioural requirements to be able to correctly advise the new owners and check on the following:

- How well do the new owners understand feline behaviour and behavioural requirements? Advice and handouts should be provided on subjects such as:
    - Diet.
    - Environmental enrichment (see Appendix 1).
    - Provision of essential resources (see Appendix 3).

- What other pets are there in the home? Could they be a risk to the cat or could the cat be a risk to them?
- If the cat has had previous outdoor access is there provision for this to continue, i.e. access to a suitable and secure outdoor space, with minimal or manageable risk from neighbouring cats, other animals or other potential threats, e.g. traffic.
- If the cat's known history indicates that it has always been an indoor cat, are the owners also happy to provide indoor litter trays and sufficient environmental enrichment if the cat is reluctant or too fearful to go outside?

Cat Sitters (Advice for Sitters and Owners)

An arrangement between a pet owner and a pet sitter, someone who stays in the home and cares for the cat while the owner is away, can be an ideal solution for both the owner and the house/pet sitter.

- The owner has continued care of their pet(s) and their home while they are away.
- The house/pet sitter often has free accommodation and many sitters combine looking after other people's pets and houses with travel.

The welfare of the pet(s) should, however, always be a primary consideration.

Being sent to a cattery can often be stressful for cats because it involves being transported away from familiarity and the sense of security associated with the home environment. The cat sitter arrangement, which allows the cat to stay in its own home, can therefore be the least stressful option for many cats. This is not always the case, however, and even if this arrangement does cause less stress for the cat than being sent to a cattery, it might still involve some degree of stress, usually connected with being apart from familiar and bonded people, having a stranger living in the house and disruption to normal routine.

For some cats, the cattery option can actually be the least stressful option. Such cats include, for example:

- Fearful and poorly socialized cats that always hide away from and/or are aggressive to visitors.
  - Some cats may hide initially but after a short while become accustomed to new people in the house. But if the cat remains fearful and/or aggressive around strangers in their territory, the pet sitter option is unlikely to improve the cat's fear and defensive behaviour and may exacerbate the problem.
  - If the cat is fearful of strangers and is allowed outdoor access, the cat may choose to 'leave home'.
  - It is unfair to expect another person to live with a cat that could cause them injury. In a cattery situation, precautions can be put in place to protect workers from attack from aggressive cats.
- Cats who have become accustomed to travelling and to visiting the same cattery on a regular and frequent basis. If a good cattery is visited repeatedly and often, the environment and carers may become familiar and comforting for the cat.

If the cat sitter option does appear to be the better option, there are a number of things that the owner and sitter can do to help limit stress for the cat, and help to encourage a good relationship between the cat and the sitter.

- It is preferable for the cat sitter to visit the home at least once before the agreed pet-sitting period, to allow initial introductions and for the cat to become familiar with the person. It can also be a good idea, if possible, for the pet sitter to arrive at the home some time before the owners depart. This can ease the transition for the cat rather than being left alone and then be faced with a less familiar person entering the house without the comforting reassurance of the owners' presence.
- The cat sitter should initially approach and interact with the cat as described in Appendix 17. This form of approach should continue until the cat is entirely comfortable with the cat sitter.
- The cat sitter should be calm and patient as it may take a while to fully gain the cat's confidence.
- The owner should inform the pet sitter, and preferably provide written information regarding the cat's normal routine, which the pet sitter should aim to keep to as much as possible. This information should include:
  - When, where and how the cat is normally fed.
  - Grooming routines – how, when, and where.
  - Litter tray information – cleaning routine, which litter substrate to use, etc.
  - Times and places where the cat and owner usually sit and relax together.
  - The rooms or areas of the house where the cat is allowed or not allowed access.
  - Whether the cat is permitted to sleep on the owners' bed at night.
  - Playtimes – when and where and the cat's favourite toys.
  - If, when and how the cat is allowed outdoor access.
  - The cat's preferred resting places and usual sleeping routine.
- The pet sitter should also be made fully aware of any current or previous health issues that might recur, and be left with contact details of the cat's usual veterinary surgeon.
- If the cat requires medication, full details should be passed on to the cat sitter, including when and how the medication is usually administered. Ideally the method of administration should be demonstrated by the owner and practised by the sitter prior to the time that the owners depart.
- Routine treatments such as flea and tick preparations, wormers, vaccinations, etc. must be up to date when the owners depart and, unless unavoidable, should not be left for the cat sitter to arrange or administer. If the owner is away for a long time so that the sitter will need to apply routine treatments, sufficient product and full instructions must be left with the sitter.
- The owner might wish to leave the house in a clean and presentable fashion for the pet sitter, but it is important that the cat's bedding is not washed prior to their departure so that it retains the cat's scent.
- Provision should be made for the cat sitter and the owner to remain in contact with each other and the cat sitter should report any concerns or changes in the cat's health or behaviour to the owner as soon as possible.

Cat Cafés

Cat cafés first became popular in the Far East, but the concept has spread around the world and there are now several across Europe and the USA.

Their popularity, especially in Japan, has been based on the beneficial effects that interacting with, or just being close to, a companion animal can have for people

Fig. 12.3. In cat cafés, and similar establishments, care should be taken that the cats are not subjected to frequent or constant unwanted attention from strangers.

(Plourde, 2014), but unfortunately, there appears to have been less interest in the welfare of the cats involved. Stress is a very real health and welfare issue for cats (see Chapter 6) and feline behaviourists and welfare experts believe that cat cafés can be highly stressful environments (Bradshaw, 2013).

Cats in cat cafés are expected to cope daily with the attentions of numerous unknown people (Fig. 12.3). This can be a significant stressor because:

- Cats need to have substantial periods of time when they are left alone so that they can sleep and engage in solitary natural behaviour. Being frequently disturbed can be highly stressful for them.
- Although most well-socialized cats do enjoy interactions with people such as play and being stroked, many cats prefer these sessions to be fairly brief and infrequent and to be able to initiate interactions with people themselves. Feeling 'in control' is very important to cats.
- Unpredictability is another major feline stressor. The behaviour of owners and people who are well known to the cat can be fairly predictable, but not so the behaviour of strangers.

But it is not just people with which cats in cat cafés need to be able to cope. Most cafés or similar establishments have several cats housed together in what is often a relatively small space. Being descended from a species that lives a primarily solitary life, the ability to be social with members of the same species can be both limited and variable between individuals. And even close-bonded relationships can easily break down, especially when the cats are faced with other stressors. In confined areas, signs that cats are not getting along can be difficult to spot because the cats are more likely to inhibit antagonistic behaviours to avoid the risk of physical conflict (see Appendix 16). But even if a stable group of cats is established, any disturbance to the group dynamic, for example the loss of a member or the introduction of a new member, illness, even short periods of separation such as vet visits, etc., can severely damage that stability.

So, although cat cafés are popular with people, they are generally not a good idea for cats. These establishments are, however, becoming increasingly popular and it seems doubtful that the trend is likely to decline soon. Consequently, I feel that it is important to provide some advice on how stress for cats in cat cafés and similar establishments might be managed and minimized as best as possible.

Seek and follow expert help and advice

Before setting up and during the running of such a venture the proprietor will need to seek help and advice from a wide range of other people. Where the welfare of the cats is concerned, ensure that advice is sought from the right people, i.e. a veterinary surgeon and/or nurse/technician with expertise and experience in feline health care, and a qualified, experienced and reputable feline behaviourist.

Sourcing the cats

- The cats must be young, healthy, neutered, vaccinated and, very importantly, well socialized with both people and other cats. Rescue cats with unknown history, stray or feral cats are best avoided as they are much less likely to be able to cope in a cat café environment.
- A group of cats that have grown up together, especially if they are related, are more likely to be bonded and get along together. But this cannot be guaranteed and even if they are closely bonded there can be no assurance that a strong bond between them will persist.
- If the cats are not to be allowed outdoor access, it must be confirmed that they have always been indoor cats and have shown no desire to venture outside. Cats that have previously been allowed to go outside can soon become frustrated and agitated if confined indoors.

Try to maintain a stable group

- As long as the cats appear to be getting along well, try not to disrupt the group, but rehome any cats that are showing signs of not getting along with the others (see Appendix 16), exhibiting aggression (towards people or other cats), or showing any signs of stress (see Appendix 18).
- Options for rehoming to suitable homes must always be available as soon as possible for cats that are unwell or showing signs of stress, aggression or other behaviour problems.
- Any new cat must be introduced to the group carefully, in the same way as introducing a new cat to an ordinary home (see Appendix 5). The new cat should not be let out into the café until it has been well accepted into the group and is itself calm and relaxed around the other cats.
- Some cat cafés have cats from local shelters that are available for rehoming. The intentions behind this may be commendable, but the constant changes to the

group and instability this creates can be highly unsettling and extremely stressful for the cats concerned. Therefore, this practice is not recommended (see An Alternative Idea later).

Staff

- Staff should have at least a good basic education in feline behaviour, health and welfare and have received their education via a reputable course provider.
- Staff must be able to recognize and report as soon as possible any signs of feline stress (see Appendix 18).
- A sufficient number of staff should be employed, not only to run the café efficiently but also to be able to closely monitor the customers' interactions with the cats, and to intervene and educate if any interactions are inappropriate or unwanted by the cat.

Design and equipment

The interior should be designed to provide the cats with unrestricted opportunity to get away from, or interact with, customers or the other cats as they wish. This should include:

- Ample access to elevated areas, e.g. plenty of shelves and walkways where the cats cannot be reached by people.
- Hiding boxes in all areas, all of which should have entrance and exit holes so that when a cat goes inside it does not get trapped there by a person or another cat.
- Easy access to rooms separate from the public space, where the cats are also provided with resting areas, elevated spaces, etc. This should also be the area where the cats are fed and have access to litter trays.
    - A good idea can be to provide exclusive individual access to areas, away from the other cats, where each cat can be fed, rest and use litter trays. These areas could be designed in a similar way as described previously for individual cattery/shelter pens and be accessed via microchip-operated cat flaps.
- Soft comfortable bedding containing the cats' scent should be supplied in a variety of locations.
- Suitable toys, such as wand toys that customers can use safely (see Appendix 2), should also be made available.
- Plenty of resources for the cats must be made available to avoid competition. A rough guideline for all resources is one per cat plus at least one extra (see Appendix 3).

Customers

- The number of customers should be limited so that the cats are not overwhelmed with too many people at one time, and staff are easily able to serve and supervise everyone. An appointment system whereby customers pre-book a timed session can be a good way to keep numbers under control.

- There should be a limit on the number of children permitted at each session. All children must be fully supervised by a responsible adult.
- On arrival customers should be informed of house rules, which should also be supplied in writing or posted in clear sight, regarding their interactions with the cats. These should include:
  - Leave the cats alone if they are sleeping, hiding or in certain areas of the café.
  - Do not attempt to pick the cats up.
  - Do not prevent the cats from going where they want to.
  - Allow the cats to come to you, rather than going to them.
  - Do not make loud or sudden noises.
  - Do not encourage the cats to play with fingers, etc. but use the toys supplied.
  - Instructions can also be supplied on how best to approach and stroke the cats (see Appendix 17).

An Alternative Idea

A more cat-friendly idea could be a cat-themed café, pub or other establishment that does not have cats on the premises. The café could be a centre for entertaining enlightenment in the form of fascinating facts and tips on cat care, history and behaviour printed on menus, coasters, posters, etc. and could even be used as a centre for talks and more in-depth education events.

If run in association with a local rescue shelter or rehoming centre, the café could also be used to publicize the work of the shelter and help boost rehoming figures by providing information, including pictures and possibly video footage of cats requiring homes. The need for volunteers and fosterers, what is involved and how to apply could also be advertised.

References

Bradshaw, J.W.S. (2013) Are Britain's cats ready for cat cafés? *Veterinary Record* 173, 554–555.

Gourkow, N. (2016a) Causes of stress and distress for cats in homing centres. In: Ellis, S. and Sparkes, A. (eds) *ISFM Guide to Feline Stress and Health. Managing Negative Emotions to Improve Feline Health and Wellbeing.* International Cat Care, Tisbury, Wiltshire, UK, pp. 82–87.

Gourkow, N. (2016b) Prevention and management of stress and distress for cats in homing centres. In: Ellis, S. and Sparkes, A. (eds) *ISFM Guide to Feline Stress and Health. Managing Negative Emotions to Improve Feline Health and Wellbeing.* International Cat Care, Tisbury, Wiltshire, UK, pp. 90–102.

Gourkow, N., LaVoy, A., Dean, G.A. and Phillips, C.J.C. (2014) Associations of behaviour with secretory immunoglobulin A and cortisol in domestic cats during their first week in an animal shelter. *Applied Animal Behaviour Science* 150, 55–64.

Halls, V. (2017) The super fosterer – how fostering cats can benefit their wellbeing. *Proceedings of the iCatCare Feline Day 2017, Birmingham, UK.*

Plourde, L. (2014) Cat cafés, affective labor, and the healing boom in Japan. *Japanese Studies* 34, 115–133.

Appendix 1
Environmental Enrichment

Cats may spend a lot of time asleep, but when they are awake they require sufficient mental and physical stimulation to keep them entertained. A cat's physical surroundings are very important; an impoverished environment that does not provide the opportunity for movement, exploration and rewarding mental challenges can lead to boredom and frustration and be a common contributory factor in the development of behaviour problems. Environmental enrichment is therefore essential, especially for cats that do not have regular outdoor access to an area large enough to allow them to hunt and explore.

Space

Cats should have sufficient space to run around. It is quite normal for a household cat to occasionally run around the house at high speed as a means of energy expenditure. But, as well as floor space, a cat also needs to have access to vertical space at varying levels, for climbing, exploring and resting. Access to shelves, high furniture and even stairs can allow this, but additional levels can also be provided in the form of cat trees (Fig. A1.1) and walkways (Fig. A1.2).

Hiding Areas

Even the most confident cat will need places to hide occasionally. But hiding places are not just used when a cat feels at threat, they can also be used in play, as areas to explore and places to retreat to when the cat does not want to be disturbed. Suitable hiding places can include gaps under furniture, covered beds and cardboard boxes. (Fig. A1.3).

Food Foraging and Food Puzzles

If we did not feed them, cats would need to hunt several times a day for their food. Having food placed in a dish in front of them may be the easy option but it can also be quite boring. It can be a lot more exciting and interesting for a cat if food is provided in way that requires some element of challenge to get it.

The easiest way to make a cat work for its food is to just scatter dried food over the floor for the cat to sniff out. But you can also buy and/or make food puzzles or food foraging toys to add in extra challenge and fun for the cat.

Fig. A1.1. A simple cat tree providing extra vertical space.

Stationary puzzles

These usually require the cat to use its paws to get the food and require a varying degree of skill and effort. Simple mazes or obstacles and food-containing compartments with sliding covers that the cat pushes or pulls open are a common feature.

Homemade versions

- Place small amounts of either wet or dry food in an ice cube tray.
- Cut holes in the sides of a plastic food container or small carboard box and place food inside. If using a plastic container, ensure that the edges of the holes are not sharp before giving it to your cat.
- Stick cardboard toilet roll or kitchen paper inners together in a rough pyramid shape and place a few pieces of dry food or cat treats inside each tube.
- Cut cardboard kitchen paper inners into varying lengths, each shorter than the cat's front legs. Then stick the tubes upright onto a base. Pour a little dry food into each tube.

Fig. A1.2. Indoor cat walkways. Similar constructions can also be provided in outdoor enclosures. Pictures courtesy of Tanya Cressey and Pause Cat Café, Bournemouth.

Appendices 197

Fig. A1.3. A cardboard box can be an easy and inexpensive way of providing environmental enrichment.

Moving puzzles

These are food toys that the cat needs to make move to get the food or treats. The simplest design is a plastic ball with at least one hole in it. Dry food and/or treats are put inside the toy which fall out as the cat pushes the ball along with his paw or nose. But there are other designs in the form of tubes or 'dangling' toys that the cat must 'bat' to get the food to release, and individual small toys containing food, that are intended for the cat to treat as pretend prey. Most of these are only suitable to be used with dried food.

Homemade versions

- Seal up the ends of a cardboard toilet or kitchen roll inner, make a few holes in the tube and then pour in few treats and/or dried food.
- Punch 1–2 holes in the bottom of an old yoghurt pot, just large enough so that small pieces of food will fall out if the pot is shaken. Attach some string to the top and the tie to a door handle or somewhere similar so that it is just above your cat's head height. To get the food your cat will need to hit the pot with his paws to get the food to fall out.
- Make holes in a small plastic pot or similar item that can be made to move freely when the cat bats or taps it with its paws.

Points to consider

- If a cat has never used a food puzzle before it is best to initially buy one that requires minimum skill or effort or one where the difficulty can be increased from an easy starting point. If the puzzle is too difficult for the cat it could cause frustration.
- The toy should not be flimsy; it should be reasonably solid and well made. Most are made of plastic, which could produce dangerous sharp edges if they break.
- Many can be used with either wet or dry food, but some are specifically designed for one or the other. Make sure that the puzzle is suitable for use with the cat's preferred food and treats.
- The puzzle should be easy to clean, especially if using wet food.

- Look at food puzzles designed for dogs as these can also be suitable for cats.
- To make each game as attractive as possible for your cat, add in a few extra tasty treats alongside the cat's everyday food.
- Safety must always be of prime importance when making home-made toys, especially those that are designed to hold food for your cat. So, make sure that there are no sharp edges and you are careful to only use safe non-toxic materials.

Reference and Further Reading

Food puzzles for cats (www.foodpuzzlesforcats.com)

© Trudi Atkinson.

This appendix is published under a Creative Commons Attribution-NonCommercial-NoDerivatives 4.0 International Licence (http://creativecommons.org/licences/by-nc-nd/4.0/)

Appendix 2
Play

Although the frequency of play activity decreases with age, it is an activity that is just as important for adult cats as it is for kittens. Insufficient opportunity for play can lead to behaviour problems such as aggression related to frustration and stress.

Cats and kittens engage in two types of play.

Social Play

This is play with another cat or kitten, or sometimes with another animal with which the cat has become well socialized. The following behaviours are a normal part of social play:

Stand-over and belly up: One cat or kitten lies on its back and 'fights' with another cat that stands over it. Claws remain retracted and bites are gentle and inhibited.
Pounce: The cat or kitten crouches and then pounces on or towards another cat or kitten. May be used to initiate play.
Chase: Running after or away from another cat or kitten.
Face-off: Sits near another cat or kitten and bats or swipes at it with retracted claws.

Social play or fighting?

Social play is the most common form of play between young kittens prior to weaning. As kittens grow up, object play becomes more common and there can be an increased risk of social play developing into real conflict. Also, because social play can appear very similar to fighting it can be very easy to misinterpret these behaviours. It is important to allow cats to play together; however, it is also important to be able to recognize the differences between social play and actual fighting.

If it is play:

- The cats are more likely to chase each other equally and swap roles.
- Claws remain retracted and bites are inhibited.
- There is no, or very little, vocalization.
- Following play the cats want to remain near to each other, groom each other or rest very close to each other.

If it is fighting:

- One cat is more likely to chase or 'pick on' the other.
- Claws are more likely to be extended and bites less inhibited.

- There is likely to be hissing, growling and/or 'shrieking'.
- If one cat is the victim, it will generally try to avoid or run away from the other cat. (This can often make the problem worse as it can trigger the chaser to chase.)

If your cats are fighting, then it is important to discover the underlying cause and address the issue correctly with professional help.

Object Play

This is play directed towards items that are pounced on, batted, chased, and/or grabbed with teeth and/or claws. It is often regarded as 'practice' of predatory skills, but because these are behaviours that cats are instinctively programmed to perform, increased play does not make a better or more prolific hunter. However, a cat that has reduced opportunity to hunt is more likely to need increased opportunity to express these normal behaviours in the form of play.

Playing with a cat or kitten can be a fun and rewarding experience for both cat and human carer but it is important to play with the cat or kitten correctly:

- **Do not** encourage a cat or young kitten to play with fingers or toes. As the kitten gets bigger and stronger more force will be applied to teeth and claws during play, this can then become very painful for the person acting as 'prey'.
- **Do not** attempt to punish a cat or kitten that might mistakenly use teeth or claws on you during play because this can lead to fear and more serious defensive aggressive behaviour. Because cats are stimulated to play by movement, the best action is to keep still, very slowly withdraw your hand or foot away from the cat and re-direct the cat's play behaviour towards a more suitable toy.
- **Do not** use a toy with hard edges or bits that could come off and be swallowed.
- **Do** use a toy that moves or can be made to move easily.

Motorized toys

These are usually battery operated and move, or have an attachment that moves in a way that mimics prey such as mice or birds. Their main advantage is that apart from switching them on they require no effort from us. But there are disadvantages:

- They can be costly in comparison to other cat toys.
- The risk of injury to your cat is slight but, even so, it is not advisable to use them when the cat cannot be supervised.
- They are not so much fun when the batteries run down or if they break.
- Some cats and kittens can be frightened of them or of the noise they make.

Bat and chase toys

A 'bat and chase' toy is anything lightweight enough that it can be made to move easily by your cat batting at it with a paw. Anything offered to your cat to play with must be safe for your cat or kitten (i.e. make sure that it does not have sharp edges or could be easily swallowed). Examples of potential bat and chase toys are:

- Table tennis balls.
- Old wine corks.
- Dried pasta shapes.
- Walnut shells.
- Balls of scrunched up paper.
- Cotton reels.

Lights, laser pens or shadows

Although these can initially seem like good fun for the cat to chase, they can cause frustration and associated behaviour issues because there is nothing for the cat to catch.

Wand toys

Sometimes called 'fishing rod toys', these usually comprise a stick, or 'wand' to which is attached a long string, sometimes elasticated, with a toy at the end. In some cases, the 'string' itself is the toy, which can be made to 'wiggle' and 'slither' along like a snake. The advantages of these toys are that even small movements of the 'wand' result in much larger movements of the toy, plus the toy that the cat uses his teeth and claws on is kept well away from your hands. (Fig. A2.1).

How to use a wand toy

- Draw the toy in long fast sweeps along the ground. This is often best to get the cat's attention towards the toy and encourage play.
- Move the toy directly upwards, or in circles in the air, to encourage leaping and grabbing.
- Move the toy in small, fast erratic movements to mimic prey.

How to Tell if the Cat is Interested in Play

Play is not limited to chasing and trying to grab a toy. For some cats just watching a moving toy can be a sufficient form of play. You can tell if a cat is interested in the toy by looking at his face. As well as obviously watching the toy move, his ears will be directed forwards and you will also see a change to his whiskers and face. The whiskers play an important role in hunting and are pushed slightly forward when the cat is in a predatory/play mood. Use of the facial muscles to move the whiskers forward also gives the cat a 'puffy cheek' appearance (Fig. A2.2).

Fig. A2.1. Playing with a 'wand toy'.

Fig. A2.2. Forward facing whiskers and 'puffy cheeks', a sign that the cat is ready to play. Note also the dilated pupils, a sign of increased arousal.

Ending the Game

Be careful not to end the game too soon. Cats will often appear to stop playing and walk away from the toy and watch it from a short distance away. Removing the toy, or stopping the game at this point may trigger frustration as this is part of the normal predatory sequence and it does not mean that the cat is no longer interested. After watching for a short time, the cat will often go back to playing; however, if the cat leaves the room completely or curls up to go to sleep, then this can be a good time to stop.

Try to avoid a 'reward crash'. This can happen if the cat has a very high motivation to play and a game ends suddenly while he is in a highly aroused 'playful' mood. If this happens the cat may redirect his predatory type play behaviour towards nearby people or other animals.

- Gradually reduce the movements of the toy, and eventually stop moving it altogether before removing it.
- If he does not lose interest, distract him onto another activity that has lower arousal potential, e.g. throw a few food treats onto the ground nearby that he has to sniff out.
- Remove the toy while he is looking away from it, e.g. while searching for or eating scattered food treats. If he sees the toy move this could trigger further play activity.

Catnip

Catnip can have highly variable effects on cats. It can often be used to enhance and encourage play. On some cats, especially young kittens, it has no effect at all, but on others it can significantly increase arousal to the point where it may even increase the likelihood of play developing into aggression. It is therefore a good idea to use catnip toys carefully until you are sure of your cat's individual reaction to the herb. Also, be aware that the amount and freshness of the catnip in cat toys can vary considerably.

© Trudi Atkinson.

This appendix is published under a Creative Commons Attribution-NonCommercial-NoDerivatives 4.0 International Licence (http://creativecommons.org/licences/by-nc-nd/4.0/)

Appendix 3
Reducing Resource Competition in a Multi-cat Household

Conflict between cats that live in the same house is most often due to competition for important resources, such as food, water, resting places and even litter trays. How we provide and locate these resources can help greatly to minimize competition and reduce the risk of fighting and other conflict behaviours.

Food

Feed household cats separately with a good distance between their food dishes

Unlike dogs and humans who are affiliative hunters and food gathers, cats are solitary self-reliant predators. In other words, each individual cat hunts only to provide food for itself, the exception being a mother cat with kittens. This also means that cats do not share their food and will choose to eat at a distance from other cats, who could potentially steal a highly valued and, for a wild or feral cat, hard-won meal.

If food dishes are placed close to each other or the cats are expected to eat from a single dish, they then will have no other option than to eat side-by-side, giving the false impression that they are happy to do so, but in fact this situation is a common cause of stress and increased antagonism in multi-cat households.

Provide extra feeding areas

It is common to provide just one food bowl for each cat, or possibly just one extra if the cats eat both wet and dry food. However, a wild or feral cat will prefer to use a few different hunting and eating locations, which not only increases the prospect of a successful hunt but also decreases the chances of conflict with a rival.

If food dishes or other important resources are located only in one area of the house it is not uncommon for one cat to sit in a doorway or corridor leading to that area and 'block' the other cats' access to the resource, creating a conflict situation between them.

It is therefore advisable to provide extra food dishes, or puzzle feeders, in more than one area of the house.

Water

Provide water dishes in various locations away from food dishes

Water is another valuable resource so a sufficient number of water dishes needs to be provided. Some cats can be reluctant to drink near their food so it can be better that water is located at a distance away from food dishes. Some cats also prefer to drink running water so provision of at least one water fountain can therefore also be a good idea.

Resting Places

Provide ample, comfortable resting places at various levels

Cats that are closely bonded will usually sleep and rest in very close contact with each other. Cats that are not closely bonded may also sleep or rest on the same item of furniture but it is more likely to be a preference for the location rather than a desire to be near each other that draws them there. If warm, comfortable resting places are in limited supply this can force cats that are not closely bonded into unwanted close contact, which can lower their tolerance of each other and result in conflict.

Litter Trays

Provide at least one litter tray per cat plus one extra, with each tray in a separate location

A toilet might not appear to us to be something worth fighting over but it is a very important resource for cats. Many cats do not like to share litter trays and it is often the case that a cat will prefer one area to urinate and a separate area to defecate. Insufficient provision of litter trays, or litter trays placed too close together, can be a common cause of conflict between cats as well as house-training issues.

© Trudi Atkinson.

This appendix is published under a Creative Commons Attribution-NonCommercial-NoDerivatives 4.0 International Licence (http://creativecommons.org/licences/by-nc-nd/4.0/)

Appendix 4
Neutering

Neutering, sometimes known as 'sterilizing', is the removal of the reproductive organs.

Spay

This is the surgical removal of the ovaries and uterus (womb) of a female cat.

Castration

This is the surgical removal of the testes of a male cat.

Why Neuter?

Population control

The most obvious reason for neutering is to avoid unwanted litters. An unneutered female cat can produce on average 2–3 litters a year with an average litter size of 3–5 kittens. This means that an unneutered queen cat could produce 15 or more kittens a year, which could add up to more than 150 kittens in an average lifetime (Robinson and Cox, 1970). An unneutered tom could be responsible for very many more, all of which would require caring homes, but sadly the majority would end up as strays, in rescue shelters or euthanized.

Health issues

- Pregnancy, giving birth and lactation are physically demanding for a female cat and repeated pregnancies can seriously affect her health.
- An unneutered female in oestrus (the time when she is sexually receptive, often known as being 'in season') may attract entire males who are less likely to be vaccinated or well cared for and therefore more likely to have infectious and potentially contagious diseases. Likewise, an unneutered male may be attracted to less well cared-for unneutered females.
- Infectious diseases and congenital disorders can be passed onto the kittens.
- As well as avoiding all of the above, neutering also greatly reduces the likely incidence of mammary tumours and pyometra (infection of the womb), both of which can be life threatening.

Control of undesirable behaviours

Urine marking (spraying)

Although neutered cats may still urine mark in response to stress, spraying is a behaviour that occurs far more commonly and frequently in unneutered males who use strong-smelling urine to mark their territory, which can include other people's property as well as your own. Entire females will also spray when in oestrus. Neutering greatly reduces the incidence of urine marking.

Vocalizing

Female cats 'call' loudly and frequently when in season. Males also vocalize loudly during the mating season when they become aware of a sexually receptive female.

Fighting

Fighting between unneutered tomcats is far more frequent, loud and intense than fights between neutered cats, and can often result in injury to one or both cats.

Roaming

Unneutered males will roam over considerable distances to find a receptive female and be at greater risk from road traffic accidents and other dangers.

When to Neuter

Kittens should be neutered before they become sexually mature. Males can be castrated as soon as the testes have descended and, to avoid pregnancy, a female should be spayed before her first season. The timing of a female's first season can vary; the majority will have their first season at around 6 months of age, which for many years has been the traditional age of neutering. However, a kitten can come into season as young as four months of age (Joyce and Yates, 2011).

Anaesthesia and surgery are potentially stressful and to minimize stress it is best that neutering is performed when it does not coincide with other stressful events such as vaccination and rehoming.

The 'Cat Group', a collection of professional organizations dedicated to feline welfare, recommend neutering of pet cats at around 4 months of age, or younger for those at greater risk of early pregnancy or when the opportunity to neuter later may be limited (http://www.thecatgroup.org.uk/policy_statements/neut.html).

References

Joyce, A. and Yates, D. (2011) Help stop teenage pregnancy! Early-age neutering in cats. *Journal of Feline Medicine and Surgery* 13, 3–10.

Robinson, R. and Cox, H.W. (1970) Reproductive performance in a cat colony over a 10-year period. *Laboratory Animals* 4, 99–112.

© Trudi Atkinson.

This appendix is published under a Creative Commons Attribution-NonCommercial-NoDerivatives 4.0 International Licence (http://creativecommons.org/licences/by-nc-nd/4.0/)

Appendix 5
Introducing an Additional Cat to your Household

Points to Consider Before Getting Another Cat

Although descended from a primarily solitary species, domestic cats have evolved the ability to be social and to enjoy and benefit from the company of other cats that they have a close bonded relationship with. However, not all cats get along well and living in the same house with another cat who may be regarded as a threat or a rival is a common source of feline stress, often resulting in behaviour problems such as fighting and indoor urine marking.

If you already have one or more cats, it is wise to be aware of the following factors that might influence the likelihood of a good or bad relationship developing between the resident cat (or cats) and a potential newcomer.

The age of the cats

The younger the cats are, the more likely it is that they will accept each other. If littermates remain together into adulthood, they will often continue to have the same close relationship they had as kittens. But even unrelated kittens have a better chance of developing and maintaining a good relationship with each other if they stay together than cats that meet for the first time when they are older. As cats become adult there is an increased chance that they will regard each other as rivals rather than as potential companions.

Early experience with other cats

Cats that had positive encounters with adult cats, not just their mother, when they were kittens (preferably between the ages of 2 and 7 weeks) are more likely to be tolerant of other cats when they grow up.

Previous experience of living with other cats

Previous positive experience with other cats can increase the chance of acceptance. But if the resident cat is grieving the recent loss of a close companion, the introduction of a possible adult rival or energetic kitten may add to, rather than alleviate, the cat's overall distress (see Appendix 6).

Introducing a kitten to an adult cat

Even if a kitten is accepting of a new adult cat that it is introduced to, a boisterous and playful kitten can sometimes be too much for, and may even frighten, an adult cat that has little experience of kittens.

The health of the cats

A cat that is elderly or unwell needs to have a stable sense of security within its core territory. Introductions to other cats at this time may heighten stress and cause increased damaged to health and general welfare.

The relationship between currently resident household cats

If you already have more than one cat at home, the state of their relationship may provide some idea of how well another cat may be accepted. An additional cat might be less likely to be accepted if the resident cats exhibit any of the following:

- Occasional fighting, even if mild.
- Indoor urine marking.
- Over-eating, by one or more of the cats.
- Blocking behaviour: sitting in doorways or corridors leading to areas where there are resources such as food, litter trays or access to the owner.

Any of the above can indicate an already fragile relationship between the cats. The introduction of another cat could precipitate further breakdown in the relationship and the escalation of behaviour problems.

Correct Introductions: Increasing the Chances that a New Cat and Resident Cat(s) will Accept Each Other

Preparation

Scent introduction

A few days before bringing your new cat or kitten home, take a cloth containing the scent of your home and cat(s) to the place where the new cat or kitten is resident, e.g. the breeder's home or rescue centre. Ask for this to be placed with your new cat. At the same time bring back home something containing the scent of the new cat or kitten for your resident cats to investigate.

Preparation of a 'safe room'

Prepare a separate 'safe' room for the new cat. This should be somewhere quiet, away from other pets (especially other cats), children, loud noises, and lively activity.

The room should contain all that the cat needs:

- Food.
- Water, placed away from the food.
- A litter tray, placed away from food and water.
- Comfortable and warm bedding.
- Toys.
- Hiding and 'safe' places, for example, cardboard boxes, access under furniture and/or high places.

The safe room should also be somewhere that the new cat will later have easy access to. Meanwhile, in readiness for when the cats are introduced, ensure that around the rest of the house there are plenty of places where the cats can hide, get up high or otherwise get away from each other (Fig. A5.1).

Bringing the new cat home

- Always transport the cat or kitten in a secure purpose-built cat carrier. Hold the carrier close to your chest rather than by the handle because the 'swinging' motion when you walk might be disorientating and unpleasant for the cat or kitten.
- A cloth or item of bedding containing scent from the cat's current home should go into the carrier with the cat.
- If transporting by car, the carrier should be well secured using a seat belt or similar, to prevent it from moving around.
- When you reach your home take the cat in the carrier straight to the pre-prepared 'safe room'.

Fig. A5.1. Ensure that there are plenty of elevated areas and hiding places around the house so that the cats are able to avoid each other if they wish.

- Open the carrier and allow the cat or kitten to exit in his own time. Don't lift the cat out or make any fuss.
- Allow the new cat to explore the room and/or hide as he wants. Don't try to restrain him or remove him from a hiding place. Give him time.
- Keep the carrier containing the 'familiar scent cloth' open and in the room with him, or if he finds another preferred hiding place put the cloth there.
- Allow the new cat or kitten to become settled and relaxed in this safe room before allowing access to the rest of the house, and before meeting your other cat(s).

Scent swapping

This allows your cats to become accustomed to each other's scent before meeting each other. It might also give you some idea of how well, or not, they are likely to get along.

- Using a clean dry cloth stroke the new cat paying particular attention to the following areas: under the chin, the side of the face and along the side of the body.
- Take the cloth to your resident cat and allow him to sniff it. Then, but only if he is happy for you to do so, stroke him with the cloth as you did with the new cat.
- Take the cloth back again to the new cat and repeat the same actions. Do this a few times daily.

Good signs: Attempting to rub against the cloth after sniffing it or no reaction. Happy to be stroked with the cloth.
Bad signs: Drawing back away from the cloth, hissing or growling. Not happy to be stroked with it. Urine spraying after having sniffed the cloth.

Room swapping

Once your new cat is settled and relaxed in his own room, try swapping areas; in other words, shut your resident cat in the new cat's room for a short time while the new cat has a chance to explore other rooms in the house. Do this over a few days.

Introductions

By now both (all) cats should be aware that there is another cat in the house and as long as everyone appears to be reasonably relaxed it should then be time for gradual 'safe' introductions.

- Start by wedging the door of the new cat's 'safe' room open sufficiently so that the cats can see and sniff at each other but are not able to get to each other and are able to retreat back to their own safe territory.
- Playing games, e.g. with wand-type toys and giving a few tasty food treats, can be an effective way for them to make good associations with each other.
- As long as the signs are good, repeat often and in other areas of the house.
- If the signs continue to be good, eventually allow them out together. Continue to play games and offer treats when they are together.

Good signs: Vertical raised tail. General relaxed body posture. Slow eye blink. Rolling. Interested in play.

Bad signs: Staring at the other cat. Ears flattened to the side or rotated backwards. Low body posture. More interest in the other cat than in play or treats. 'Swishing' tail. Growling. Hissing. Attempting to get away. Urine spraying after meeting the other cat.

Keeping the Peace

After initial introductions, if all goes well, the cats may not necessarily develop a close bonded relationship, where they seek out each other's company and sleep wrapped around each other, but more commonly they may have a relationship where they are tolerant of each other and can live together with minimum stress or conflict. The following advice can help to support and maintain this relationship:

- Ensure that the new cat still has access to his safe room and continues to be fed and have access to all other resources in the safe room for as long as necessary.
- Make sure that there are always lots of easily accessible hiding places or high places around the house that the cats can retreat to if feeling threatened.
- Ensure that there are always adequate resources for all the cats in the household.
  - Feed the cats in separate areas and supply additional food dishes in other areas.
  - Provide extra water dishes in areas away from food.
  - Cats can be reluctant to share litter trays and many prefer one area to urinate and another to defecate. The general rule regarding the number of litter trays to provide is one per cat plus one extra. The litter trays need to be located some distance away from each other. Two or more trays placed side by side will be regarded by the cats as one litter location.
  - Ensure that they all have access to plenty of warm and comfortable resting places so that they are not forced to share sleeping areas, which might be stressful for them, unless they are closely bonded.
  - Ensure that they have plenty to do to keep them entertained and allowed to engage in normal feline behaviour. For example: interactive games, such as wand toys, and things to chase, bat and grab; food foraging toys, such as food balls and, if possible, puzzle feeders; and outdoor access (see also Appendices 1 and 2).

© Trudi Atkinson.

This appendix is published under a Creative Commons Attribution-NonCommercial-NoDerivatives 4.0 International Licence (http://creativecommons.org/licences/by-nc-nd/4.0/)

Appendix 6
Helping a Grieving Cat

Do Cats Grieve?

Following the loss of a feline companion or a family member it is sometimes reported by owners that their cats exhibit behaviours such as increased vocalizing, pacing and searching for the lost individual, followed by a reduction in appetite and becoming withdrawn and inactive. This reaction, which may last for just a few days or sometimes months, does seem to indicate that the cat may be grieving.

How Can We Help?

Getting another cat might not be the best idea

If a close feline companion has been lost, providing another 'friend' can seem to be the obvious solution. But unfortunately, this is not always a good idea and may add to, rather than reduce, the stress and anxiety experienced by the grieving cat for the following reasons:

- Despite having lived with another cat, the resident cat may not be so accepting of a stranger, especially whilst grieving for the lost companion.
- If the newcomer to the household is another adult it is possible that he may regard the resident grieving cat as a rival and act aggressively towards him.
- A kitten may be more accepting of your adult cat but a lively and boisterous kitten can often be too much for a grieving cat to cope with.

It can be far better to help the grieving cat in other ways and consider the introduction of a companion later, when he may be more able to cope.

Consult your vet

It is especially important to contact your vet if your cat is off his food or lethargic for more than 24 hours because these symptoms might not just be due to grief, but could indicate that your cat is unwell.

Try to keep to your cat's normal routine

Routine and predictability are very important for cats. The loss of the companion will have already resulted in some changes to normal circumstances, so it is important to make sure that everything else remains the same.

Don't throw away the other cat's bedding, etc.

Bedding, clothing, etc. that contains the scent of the individual who is no longer around can provide the remaining cat with some comfort, and the gradually fading scent can help the cat to cope and confirm the absence of the companion.

Be available for your cat

It is important to be around for your cat, especially when he wants comfort or attention from you, but always allow him to come to you rather than forcing your attention on him as this could also be stressful for him and at times he may prefer to be left alone.

Other cats

As well as grief at the loss of a companion, it is also important to be aware of other possible effects that losing a cat can have on other cats that live in the same household.

Other household cats

In a multi-cat household the loss of one can sometimes disrupt the relationship between the others. This situation is often short-lived and the cats will usually sort things out for themselves within a few weeks. But if serious fighting occurs or if the situation does not appear to be improving, it can be best to seek professional help, initially from your veterinary surgeon to check that underlying disease is not a contributory factor, and then from a suitably qualified feline behaviourist via veterinary referral.

Neighbouring cats

If your cats are allowed outdoor access, the individual that has been lost may have played an effective role in defending the territory. Now that individual is no longer around, other cats might be more likely to enter the territory and cause problems for the remaining cat(s). If this occurs it is advisable to ensure that neighbouring cats cannot enter the house and that within the immediate territory, i.e. the garden, there is ample provision for the resident cats to access places to hide, places to get up high and be above the other cats, and escape routes so that they can get away from other cats.

Further Reading

Carney S. and Halls V. (2016) Feline bereavement. In: *Caring for an Elderly Cat.* Vet Professionals, pp 113–117.

© Trudi Atkinson.

This appendix is published under a Creative Commons Attribution-NonCommercial-NoDerivatives 4.0 International Licence (http://creativecommons.org/licences/by-nc-nd/4.0/)

Appendix 7
Introducing Cats and Dogs

There are many households where cats and dogs live together peacefully; however, this is not always the case, and situations can occur that are highly stressful and even potentially dangerous for the cats involved (Fig. A7.1). With careful consideration before acquiring a new pet, and with careful preparation and introductions, the chances of such situations arising may be lessened.

If You Have a Dog and are Considering Getting a Cat

First, consider if your dog is likely to be 'cat friendly'. Unfortunately, there is no way to be completely certain that your dog will not attempt to chase or attack a new cat or kitten in the house, but your dog's past experiences with cats and aspects of his general behaviour can provide some indication as to whether or not your dog is suitable to be around cats.

Did your dog live with a cat as a puppy? Dogs that have been socialized with cats when young, preferably under 12 weeks of age, are more likely to consider cats as friends when adult.

Has your dog lived with cats previously? How was their relationship? If the dog chased the cats he lived with before, the chances are that he will also chase a new cat or kitten. However, even if you already have a cat, or your dog has lived peacefully with a cat before, it is still no guarantee that he will not attempt to chase a new cat or kitten that is introduced into the home.

Does your dog chase cats, rabbits or squirrels when on walks? A dog that chases cats and/or wildlife may also be more likely to chase a pet cat in the home. This can be of most concern if the dog actually catches and kills small animals. If this is the case it would be unwise to attempt to bring a cat or kitten into the same household.

If You Have a Cat and are Considering Getting a Dog

Some cats will be better able to accept a new dog or puppy in the house than others. Most cats regard dogs as natural predators, however, how fearful the cat is, and its reactions around dogs can depend on its previous experiences and the behaviour of the dog.

Did your cat live with a dog as a young kitten? Cats that have been socialized with dogs when young, preferably under 7 weeks of age, are more likely to consider a similar sized or breed of dog as a friend when adult.

What is your cat's previous experience of dogs? Has he lived with a dog before? If so how was their relationship?

Fig. A7.1. Although cats and dogs can live together in harmony, it is always wise to be aware that dogs can regard cats as playthings or even potential prey, and situations can occur that are potentially dangerous for the cats involved.

Is your cat fearful or aggressive around dogs? A cat that is frightened of all dogs, regardless of the dog's behaviour, is less likely to accept a dog into its home and is likely to suffer severe chronic stress if subjected to sharing its home with a dog.

Introducing a New Cat or Kitten into a Home with a Resident Dog

Choosing your cat or kitten

Choose a cat or kitten that has had positive experience with dogs of a similar breed or type as your own dog. Because of the extreme differences between dog breeds, a cat or kitten that has had good experience of one breed or type may not be as accepting of other dogs of a different size or appearance. Early experience (i.e. prior to 7 weeks of age) is of most importance; however, if you are acquiring an adult cat, the effects of this experience will be lessened if the cat has had very little or no experience of dogs since then.

Preparation

Teach your dog a 'look at me' command

It is a good idea to start to teach your dog this command long before bringing your new cat or kitten home, so that that he is already responding well by the time you first introduce your dog to your new cat.

- Have a pot of small, tasty food treats ready. It is best to start teaching the command while you are sitting down so that your dog does not have too far to look up at you.
- Wait for your dog to look at you or if necessary lure him by moving your hand up towards your eyes.
- As soon as he looks towards you, **instantly** 'mark' the good behaviour by using a clicker or by saying 'Yesss' and then give him a treat.
- Do not hold the treat ready in your hand; reward him with a treat from the pot. If during training you always have a treat ready in your hand, your dog may not respond later when you don't have a treat in your hand.
- Repeat until he starts to look towards you readily and is obviously expecting a treat for doing so!
- Start to bring in your command word as he turns to look at you. This must be a word that you would be most likely to use when you really need it. It should also be something that can be said in a clear, bright and encouraging voice, such as 'Watch' or 'Look'.
- Repeat so that your dog starts to associate the word with the action of looking towards you.
- During the training sessions, use the command to get him to look at you when he looks away.
- The ultimate aim is to get your dog to look towards you and away from the cat or kitten but we need to build up distractions gradually.

Building up distractions

- Put some fairly unexciting household object onto the floor close to you. As your dog approaches to investigate use your chosen command. As soon as he looks at you reward with a click or a 'Yesss' and then a treat.
- Repeat but with the same item further away.
- Repeat with a variety of different objects starting with boring items and gradually more interesting items, such as a favourite toy. Also do this in different places around the house and garden until the dog gets the idea that the most rewarding thing to is to look away from the item and look to you instead.
- Then try getting his attention away from 'bigger' distractions such as a favourite toy or an 'interesting person' who has just walked into the room.

Scent introduction

A few days before bringing your new cat or kitten home take a cloth containing the scent of your home and dog to the place where the cat or kitten is resident, e.g. the breeder's home or rescue centre. Ask for this to be placed with the cat or kitten. At the same time bring back home something containing the scent of the new cat or kitten for your dog to investigate and become accustomed to.

Assess your dog's reaction when introducing the scent of the cat. Ideally the dog should remain calm and relaxed and not become agitated or excited.

Preparation of a 'safe room'

Prepare a separate 'safe' room for the cat or kitten. This should be somewhere quiet, away from all other pets, children, loud noises and lively activity.

The room should contain all that the cat needs:

- Food.
- Water, placed away from the food.
- Litter trays, placed away from food and water.
- Comfortable and warm bedding.
- Toys.
- Hiding and 'safe' places, for example cardboard boxes, access under furniture and/or high places.

The safe room should also be somewhere that the cat or kitten will later have easy access to, but the dog will not. This can be achieved by using baby gates or by fitting a cat flap in the door that only the cat can use.

Meanwhile, in readiness for when the cat is allowed out of his 'safe room', ensure that around the rest of the house there are plenty of elevated places and escape routes that allow the cat to be able to get away from the dog or be out of his reach (Fig. A7.2). Make use of baby gates or similar so that the cat can access areas of the house that the dog cannot.

Bringing the new cat home

- Always transport the cat or kitten in a secure purpose-built cat carrier. Hold the carrier close to your chest rather than by the handle, as the 'swinging' motion when you walk may be disorientating and unpleasant for the cat or kitten.

Fig. A7.2. Make sure that there are plenty of areas around the house that the cat can access but the dog cannot.

- A cloth or item of bedding containing scent from the cat's current home should go into the carrier with the cat.
- If transporting by car, the carrier should be well secured using a seat belt or similar, to prevent it from moving around.
- When you reach your home take the cat in the carrier straight to the pre-prepared 'safe room'.
- Open the carrier and allow the cat or kitten to exit in his own time. Don't lift the cat out or make any fuss.
- Allow the new cat to explore the room and/or hide as he wants. Don't try to restrain him or remove him from a hiding place. Give him time.
- Keep the carrier containing the 'familiar scent cloth' open and in the room with him, or if he finds another preferred hiding place put the cloth there.
- Allow the new cat or kitten to become settled and relaxed in this safe room before allowing access to the rest of the house, and before meeting your dog. This may take a few days.
- Once the cat or kitten is settled and relaxed in his 'safe room' allow him out to explore other rooms of the house without the dog present. The best times to do this can be when the dog is out for a walk or securely shut away elsewhere. The cat should be allowed to become familiar with elevated areas, escape routes and hiding places, so that he is aware of where he can escape to if necessary when first introduced to the dog.

Introducing the dog and cat to each other

- Do not attempt introductions until the dog is responding well to a 'look at me' command and the cat is comfortable in the room where the introductions are to take place (this should **not** be the cat's safe room).
- Before introducing the dog, the cat should be allowed to settle in the room on an elevated area, from where he can feel safe and look down on the dog. You may need to encourage him up onto a suitable place using food treats.
- Bring the dog into the room on a lead that should be securely attached to a normal, comfortable flat collar or harness. The lead should not be held uncomfortably tight; it should only be used as a safeguard should the dog attempt to chase the cat. At no point should any attempt be made to punish the dog or pull him back harshly on the lead.
- Reward the dog for calm and relaxed behaviour while the cat is in the room, and use the 'look at me' command to re-direct the dog's attention towards you if he appears to be becoming more intently interested in the cat.
- At all times allow the cat to hide or escape if he so wishes and at no time attempt to restrain the cat or force him to get closer to the dog.
- Repeat until you feel that both are sufficiently relaxed in each other's company.

Introducing a Dog or Puppy into a Home with a Resident Cat

Choosing your dog or puppy

Choose a dog or puppy that has already lived with cats and got along well with them, without chasing them. Preferably choose a dog or puppy that has been well socialized with cats prior to 12 weeks of age.

Preparation

- Before bringing the dog or puppy home take a cloth containing your cat's scent to the dog or puppy and assess his reaction. Ideally the dog should remain calm and relaxed and not become agitated or excited.
- Also a few days before bringing your new dog or puppy home, introduce a cloth containing the scent of the new dog or puppy for your cat to investigate.
- In all areas of the house where the dog will be allowed, ensure that there are plenty of places where the cat can hide, get up high or otherwise get away from the dog. Make use of baby gates or similar so that the cat can access some areas of the house that the dog cannot.

After bringing the dog or puppy home

- Initially keep the dog or puppy and cat separate. The cat must continue to have access to most of its normal areas for feeding, sleeping, litter trays and access to outside if this is usual for the cat.
- Teach the dog or puppy a 'look at me' command and introduce the dog and cat as described under 'introducing a new cat or kitten'.

Using a Crate

It is sometimes recommended to use a crate, cage or indoor kennel to introduce pets. Confining a dog or puppy to a crate, while the cat is allowed free access and to become accustomed to the dog, can work well as long as the dog or puppy is used to being in a crate and will relax when confined. If not, then this could increase the dog's agitation and be potentially damaging to the relationship between the pets.

A cat must never be confined in a cage while a dog or any animal that could be perceived as a threat is allowed to approach the crate. If the cat is confined and unable to escape this is more likely to increase its fear of the dog.

© Trudi Atkinson.

This appendix is published under a Creative Commons Attribution-NonCommercial-NoDerivatives 4.0 International Licence (http://creativecommons.org/licences/by-nc-nd/4.0/)

Appendix 8
Cat Flaps

Keeping a window slightly open is one way to allow your cat the freedom to come and go as he pleases but this could be a potential security risk. Another option is that you act as doorman, opening and closing the door whenever your cat wants to go out or come back indoors. But as well as being inconvenient for you, this could also be frustrating, even stressful, for your cat, especially if you are not around when he needs to go out to eliminate or to escape from threats or stressors, either within the home or outside. Fitting a cat flap can therefore be a good idea for both you and your cat.

But because cat flaps can potentially allow other cats into the house they can also be a source of stress for your cat. Where a cat flap is located, and the type of cat flap fitted, can make this much less of an issue.

Where to Fit a Cat Flap

- Make sure that your cat has easy access to the cat flap from both indoors and outdoors. Your cat maybe agile now, but in years to come climbing or leaping up to reach the cat flap might be uncomfortable or difficult for him. Also, your cat might need to get through the flap quickly and easily if he is being chased or is frightened by something outside. If the cat flap can only be fitted somewhere off the ground, provide a ramp or steps for easy access.
- Fit the cat flap well away from the things that are important to your cat, such as food, water, and resting places. Food, or even a comfortable bed near the cat flap, may encourage other cats in through the cat flap and having important resources close to the cat flap, which will be perceived by your cat as a potential entry point for an intruder, can be very stressful for your cat.
- Litter trays should also be positioned well away from a cat flap. A cat can feel very vulnerable when using a litter tray and may be reluctant to use one that is located near to an entrance or exit, such as a doorway or cat flap.
- When venturing outside, a cat is leaving the safety and security of home and going out into a world where there could be dangers, such as other cats or neighbourhood dogs. A cat can therefore sometimes feel vulnerable when going outside through the cat flap, especially if he is going from a dimly lit environment into bright sunshine or from bright artificial light to darkness and thus might not be able to see very well for a few minutes. Providing him with somewhere to hide before venturing further can help to make him feel more secure. A good way to do this is to position bushy plants or similar close by the cat flap. Some plants can be toxic to cats, however, so it very important to ensure that the plants you choose are safe for your cat or kitten (information on potentially harmful indoor and outdoor plants can be accessed via International Cat Care: https://icatcare.org/advice/poisonous-plants).

Be wary of placing large solid objects nearby because these can provide a vantage point for other cats from which to ambush your cat as he comes out through the cat flap.

Types of Cat Flap

The simplest type of cat flap is just a flap that your cat can push through (although some cats prefer to pull the flap open with a paw!). These are of course the easiest and cheapest to install; however, they do not prevent other cats from coming into your (and your cat's) home. So you may wish to fit an 'exclusive' cat flap, i.e. one that can only be used by your cat(s).

Magnetic or infra-red operated

Advantages

- Relatively inexpensive and easy to fit.

Disadvantages

- Your cat will need to wear a collar that could cause injury to your cat if he gets caught up by it, or it could come off and be lost outside, thereby preventing your cat from using the cat flap and getting back in.
- The infra-red or magnetic locking devices are often not sufficient to prevent other cats from forcing their way in if they are strong and determined enough.

Microchip activated

Advantages

- Your cat does not need to wear a collar. The cat flap works by reading a microchip embedded under your cat's skin.
- The locking device is usually stronger than infra-red or magnetic locking devices.
- Recent designs allow you to restrict the cat to one-way access for a set period of time as required and if you have more than one cat you can choose individual settings for each cat.

Disadvantages

- Often more expensive and depending on design may be a little more complicated to fit.
- Can only work if your cat has been microchipped (although some manufacturers can also supply a collar to activate the device).

- A microchip can sometimes migrate within the cat's body to a place where it cannot be picked up by the cat flap microchip reader, but this is a fairly rare occurrence. However, before purchasing a microchip-activated cat flap it can be a good idea to ask your vet or vet nurse/technician to check that your cat's microchip is still where it should be and can be read easily.

Training Your Cat to Use a Cat Flap

You cannot fit a cat flap and expect your cat to use it straight away. He will need to learn what it is and how to use it. The simplest way to do this is to either prop open or remove the flap if possible, leaving a hole through which your cat can climb in and out. Then spend some time encouraging him back and forth, using a toy or some of his favourite treats and plenty of praise. At no time force him through the gap as this will only deter him from using it. Leave it propped open and give him time to get used to it.

Once he is fully confident going in and out through the open gap, drop down/replace the flap, but, if possible, with any locking mechanism switched off so that it can be opened without waiting for the collar or microchip to be read. Then once again encourage him back and forth with treats and praise. Once he is regularly going in and out by pushing open the flap, allow him a few days to become fully accustomed to it before activating the locking mechanism.

© Trudi Atkinson.

This appendix is published under a Creative Commons Attribution-NonCommercial-NoDerivatives 4.0 International Licence (http://creativecommons.org/licences/by-nc-nd/4.0/)

Appendix 9
Teaching Your Cat to Come to You When You Call

Recall is something we normally associate with dogs. But teaching your cat to come to you when you call is something that is not only possible, and usually quite easy to do, but it can also be very useful.

1. Experiment to find the food treats that your cat most enjoys.
2. Get a few of his favourite food treats ready in your hand or pocket.
3. While he is in the same room and not too far away from you, make a specific sound to get his attention – you can use your voice and call his name, but remember that when you need to call him he may be some distance away so a more distinct and louder sound, such as a whistle may be better, although when the cat is close by keep the volume low or else you might frighten him away, rather than encourage him towards you.
4. As soon as he comes to you after you have made the sound, give him a tasty food treat or whatever he most enjoys, for some cats this may be being stroked or playing a game.
5. Repeat in different places around the house and at different times.
6. Bit by bit increase your distance from your cat so that eventually you can call him from one room to another.
7. Eventually when he hears the specific sound, he will learn to expect a treat from you, which should encourage him to come to you from wherever he is in the house.
8. If your cat is allowed outside, you can then start to use the 'recall' to call him indoors. Start when he is still quite close to home. Repeat a few times and then try calling him when he is a little further away.
9. Eventually, as long as he can hear you, you should be able to call him back indoors from wherever he is.
10. Initially he should get a treat every time he comes to you, but once he is starting to respond well each time, it is better that rewards are given intermittently rather than every time, and the quality and amount of the reward should also vary. This will keep him keen to respond whenever he hears you call.
11. It is also important that your cat is happy to approach you at all other times, not just when you call him. See Appendix 17 and watch the following video from International Cat Care: https://m.youtube.com/watch?v=eqUpsyAiNn4&feature=youtu.be

© Trudi Atkinson.

This appendix is published under a Creative Commons Attribution-NonCommercial-NoDerivatives 4.0 International Licence (http://creativecommons.org/licences/by-nc-nd/4.0/)

Appendix 10
Cats, Babies and Children

Cats can be loving and fulfilling pets for both adults and children, but problems can sometimes arise that might result in a child being bitten or scratched and the cat becoming stressed and frightened. As well as aiming to keep the child or baby safe, it is also wise to do the best you can to avoid the cat becoming stressed, because a stressed cat is more likely to develop behaviour problems such as indoor urine marking or aggression.

A New Baby

Before the baby arrives

The arrival of a new baby can be a highly stressful time for a pet cat and it is important to prepare well in advance.

Try to keep your cat(s) out of the room that you intend to use as the nursery. It will be easier if you start to deny your cat access to this room long before the baby is due rather than trying to shut him out once the baby has arrived.
Allow your cat to become accustomed to the sights, sounds and smells of a new baby.

- Bring the equipment you will need for the new baby (buggy, carry cot, high-chair, etc.) into the house well before the baby arrives. Allow your cat to sniff, examine and become accustomed to them. Once your cat appears to accept them as he does the rest of the furniture in the house, then start to move them around the house on a regular basis, just as you are likely to do once the baby has arrived.
- The sounds of a new baby can sometimes be disturbing for a cat; it can therefore be a good idea to allow your cat to become accustomed to baby noises by playing recorded sounds at a very low level and then very gradually increasing them. Suitable recorded sounds are available for free online (for example https://soundcloud.com/dogstrust/sounds-soothing-baby-crying).
- If the baby is born in hospital, once he or she has arrived bring home a small blanket or similar that the baby has been wrapped in or slept on, and then place it somewhere that the cat can sniff it and become accustomed to the smell.
- Installing a facial pheromone diffuser at home at least 2–3 weeks before the baby arrives might also help your cat to cope.

Once the baby has arrived

Never leave the baby alone with the cat. Do not allow the cat to sleep in the same room as your baby unattended. It is an old superstition that cats deliberately suffocate babies in their sleep but cats can be attracted to warmth and a cat may lie very close to the baby's face. Also, cat dander (shed from the coat and skin) can occasionally cause a serious allergic respiratory reaction (Herre *et al.*, 2013).

Try to keep to your cat's normal routine as much as possible. Household routines are bound to change once the baby has arrived, but it can be difficult for your cat to understand this.

- Try to continue feeding your cat when and where he is accustomed to being fed.
- Continue to allow him access to familiar, safe and comfortable resting places.
- As much as you can, continue to play, fuss and interact with your cat as you did before the baby arrived.

Cats and Children

- Make sure that your cat has somewhere safe to retreat to away from the child(ren). For example:
    - Provide access to high shelves or the tops of cupboards, etc., or provide a tall and sturdy 'cat tree'.
    - Use baby gates to provide an area for the cat that he can access but the child(ren) cannot.
- Especially make sure that the child(ren) cannot disturb the cat while he is eating or sleeping.
- Prevent the child(ren) from trying to follow or disturb the cat by redirecting them onto another 'fun' activity away from the cat.
- Teach the child how to interact with the cat appropriately (Fig. A10.1). You might need to teach the child how to stroke the cat 'nicely' by gently guiding the child's hand with your own.

Fig. A10.1. Children should be taught how to stroke the cat gently and appropriately.

- If the cat is happy to be picked up, teach the child how to correctly hold the cat by supporting it fully underneath.
- If the cat does not like being picked up, dissuade the child from attempting to handle the cat. But try not to become angry with the child, try to redirect him or her towards a toy or other 'fun' activity instead.
- Never shout at, hit or try to physically punish the cat at any time, but especially not in front of the child(ren). Children often imitate the actions of grown-ups and the cat might react aggressively if the child tries to copy you. Also, the cat may associate your anger and the unpleasant event, not with his or her own behaviour, but with the presence of the child, thereby making the child appear to be a threat to the cat.
- Don't encourage your cat or kitten to play at 'attacking' your hands or feet. If the cat tries to play the same rough games with a small child the child may be hurt.
- Teach the child(ren) to always go to you if the cat does anything that they don't like. A young child may not deal with the situation in the most appropriate and safe manner.
- Your cat is likely to be less tolerant if he is unwell, so try to keep him healthy. Take him for regular veterinary checks and make sure that vaccinations, worming and flea treatments are kept up to date.
- Watch for signs that the cat may be feeling uncomfortable while the child(ren) are close by. For example:
  - Swishing or slapping tail.
  - Ears held back or to the side.
  - Low, slinking body posture.
  - Hissing.
  - Growling.
  - Trying to get away.

If you see any of these signs encourage the child(ren) away from the cat.

Reference

Herre, J., Grönlund, H., Brooks, H., Hopkins, L., Waggoner, L., Murton, B., Gangloff, M., Opaleye, O., Chivers, E.R., Fitzgerald, K., Gay, N., Monie, T. and Bryant, C. (2013) Allergens as immunomodulatory proteins: the cat dander protein Fel d 1 enhances TLR activation by lipid ligands. *Journal of Immunology* 191, 1529–1535.

© Trudi Atkinson.

This appendix is published under a Creative Commons Attribution-NonCommercial-NoDerivatives 4.0 International Licence (http://creativecommons.org/licences/by-nc-nd/4.0/)

Appendix 11
House-training Your Cat or Kitten

Kittens learn to use a litter tray from observation of their mother's behaviour, so that by the time they reach their new homes most kittens are already house-trained and, as long as they are provided with a suitable litter tray and litter substrate, no further house-training is required.

But occasionally a cat or kitten may use some other area of the house as a toilet. The most common reasons for this are described below.

Insufficient Number of Litter Trays

Cats often prefer one area to urinate and a separate area to defecate. They can also be reluctant to share a litter tray with other cats. For these reasons the recommended number of litter trays to provide is at least two for a single cat and one per cat plus one extra in a multi-cat household.

The Location of the Litter Trays

For a cat or kitten that is new to the house, it is important that the litter trays are easily accessible and easy to find. Providing separate litter trays also means providing them in separate locations. If they are placed side-by-side the cat(s) may regard this as a single litter location and still eliminate elsewhere.

Cats can also be reluctant to use a litter tray that is positioned:

- **Near food or water.** Most cats do not like to urinate or defecate close to where they eat or drink.
- **Near windows or glass doors.** A cat can feel vulnerable when he uses a litter tray and feel particularly at threat if he thinks that he may be seen by neighbouring cats.
- **Near the cat flap or other entrances and exits.** Where the cat enters or exits the house can also be viewed as a potential entry point for rivals from outside. Therefore, this is not a place where a cat is likely to feel comfortable using a toilet.
- **Near to other potential threats or disturbances**: for example, in a walkway, by the dog's bed, by a cupboard that is frequently opened, where children play, near the washing machine or other noisy household appliances, or in any area where the cat has had a frightening or painful experience. If a cat feels worried about using a tray in an 'exposed' position he may prefer to use a quieter 'hideaway' such as behind a chair, under a table or in a quiet corner.

If your cat is choosing to urinate or defecate somewhere other than where you have placed the litter trays, he may be telling you that this is his preferred location.

Therefore, placing a litter tray where he is choosing to eliminate may encourage him to use it. If the cat does then use the litter tray but it is not an ideal location for you, it can be very gradually (a few inches a day) moved to a more appropriate place.

The Size and Shape of the Litter Tray

- A litter tray should be big enough to allow the cat to turn around without any part of his body, including his tail, touching the sides, and deep enough to contain a sufficient depth of litter, 5–10cm (2–4 inches), depending on the size of the cat. However, consideration should be made for kittens, small or elderly cats who may have difficulty getting in and out of a litter tray that is too big or too high-sided.
- An elderly cat with arthritis or a cat with any condition that might cause difficulty with movement may benefit from being provided with a litter tray that has one side cut away to allow easy access. A potting tray or garden tidy with a high back and low front can also be used.

Covered vs Uncovered Litter Tray?

Whether the litter tray is covered or uncovered can be a matter of individual preference and the cat's previous experience. Some cats feel more secure using a covered tray, whereas others refuse to use them. There are a few points to be aware of if providing a covered litter tray:

- **Keep it clean.** Because a covered tray blocks the sight and smell from us, it is easy to be unaware that the tray needs cleaning. But keeping a covered tray clean and fresh is especially important because the trapped odours can soon become too strong for the cat and deter him from using it.
- **Is it big enough?** Because cats do not like to have any part of their body touching the sides of the tray while they use it, most will not like to use a covered tray that is too small. For cats with arthritis or similar conditions, this can be especially uncomfortable for them.
- **Watch out for ambushing.** When a cat is exiting a covered litter tray this can provide an opportunity for a rival housemate to ambush and 'attack'. If this is happening it can soon deter the victim from using the covered tray.

The 'Wrong' Type of Litter Substrate

The litter substrate is what the cat digs into to deposit its waste. What the cat will use can also be a matter of individual preference and previous experience.

- **Novel or inconsistent litter substrate:** A cat may not want to use a litter substrate that is different in texture or scent to one it is accustomed to.
- **Unclean:** Some cats can be more fastidious than others, but a dirty litter tray can deter many cats from using it. The best policy is to clean out any faeces and/or wet patches at least once or twice daily and wash the tray completely once or twice a week.
- **Scented:** Cats have a very sensitive sense of smell and scents that seem attractive to us can often be overpowering for a cat. Using a scented cat litter or one that releases

a scent when it is wet can be highly repellent for cats, especially in a covered tray that traps the odour. The same applies to scented tray liners, air freshening sprays or other scented products or devices used close to the litter tray.

- **Too shallow**: Most cats like to dig a reasonable-sized hole in which to urinate or defecate. They can't do this if an insufficient depth of litter is provided.
- **Uncomfortable**: The litter needs to be easy to dig into and to walk on. This is especially important for an elderly or injured cat who may find walking on and trying to dig into pelleted litter or similar difficult or uncomfortable.
- The use of plastic tray liners can also deter a cat from using the tray because while attempting to dig, the cat's claws can get snagged in the liner.

Using an 'Outdoor' Toilet

If your cat is allowed access outside you might prefer him to eliminate outside rather than in the house. It can often be best to give the cat the option of using an indoor or an outdoor toilet.

Encouraging your cat or kitten to eliminate outside

- Your kitten or cat will have learnt to eliminate on cat litter so you may need to help your cat to make the change from litter to soil. You can do this by mixing a little soil from outside into the litter tray, gradually adding just a little at a time.
- At the same time choose an area of your land that you would prefer your cat to use and empty a little of the soiled litter over that area so that you transfer the scent to outside. Cats are more likely to use an area away from the house where they feel partially hidden, so the best place may be amongst bushes, but don't be surprised if your cat chooses a different location.
- Your cat might be reluctant to go to the toilet outside if he feels vulnerable, and to a kitten the big outdoors can seem a daunting place, so until he feels confident about going outdoors he may be happier using a litter tray indoors occasionally for a little while longer. So don't be tempted to remove indoor litter trays until you are sure that he is completely at ease about going outside. If you take the tray away too soon you may find that he has found another, not so suitable, place to use as an indoor toilet!
- Never get rid of the litter trays completely as you will need them for times when your cat needs to be confined indoors e.g. whenever there might be fireworks outside, or if your cat is unwell.
- Providing a purpose-built outdoor latrine can also help to encourage your cat to eliminate outside. It can also be easier for you to keep clean and can reduce the risk of other areas of your garden being used as a cat toilet.
    - The best location is usually at the edge of the garden, surrounded by bushes, somewhere that the cat can access easily but provides some privacy.
    - Dig a hole that is slightly larger in diameter than an average litter tray, approximately 45–60cm deep, and line with plastic.
    - Fill a third with pea gravel and top up with playground sand.
    - Remove soiled patches frequently (once a day) and top up with fresh sand as necessary.

House-training 'Accidents'

If your cat does urinate or defecate in the house away from the litter tray, it is important to deal with this correctly.

Do not attempt to punish the cat. Attempting to punish your cat after the event is pointless because he will not be able to make the association between your 'correction' and the act of eliminating. Catching the cat 'in the act' can also be counter-productive because the cat is more likely to associate your presence when eliminating as the thing to avoid, which can result in the cat finding a quiet corner well away from you to use as a toilet rather than use the litter tray if you are nearby.

Clean the area: Cats may re-use areas that they have previously used as a toilet and can be drawn back to the location by the residual scent of urine or faeces.

- Wash the area initially with plenty of warm water.
- Test a small area first and then wash with a 10–20% solution of biological detergent or a proprietary enzymatic 'odour elimination' product (avoid using bleach or products containing bleach).
- Rinse, pat dry.
- Wipe or spray over with surgical spirit (again test an area first).

Unfortunately, because urine can soak into carpets, rugs, other porous surfaces and soft furnishings, cleaning sufficiently is not always possible, and it can sometimes be necessary to remove heavily soiled items or floor coverings, and deny the cat access to the area until it can be cleaned thoroughly. If it is a carpet that needs to be taken up, remove any underlay as well, clean the floor underneath thoroughly, as described above, and then leave the area uncovered for at least a week to allow it to 'air'. When replacing flooring it is often best to choose a new floor covering that has a very different texture to the previous one because the cat might have developed a preference for toileting on the previous covering and materials of a similar texture.

If the Problem Persists

If all the points mentioned above have been addressed and your cat continues to urinate or defecate in inappropriate places, it is then necessary to seek professional help.

House-training problems can often be linked to underlying health issues so it is important to get your cat checked by your vet. If your vet can find no medical problem or any relevant health issues have been treated but the house-training problem persists, then ask your vet if he is able to refer you to a qualified behaviourist with suitable knowledge and experience of feline behaviour.

House-Training Problem or Scent-Marking?

Cats may also deposit urine around the house as a way of scent-marking. The main difference between a cat that is urinating to relieve itself and one that is using urine to scent-mark (a behaviour known as 'spraying') is the position that the cat adopts. A cat that is relieving itself will squat down on its hind quarters and hold its tail horizontally out behind it. Whereas a spraying cat squirts urine backwards whilst standing, often on tiptoes and with the tail raised vertically. Urine spraying is a normal part of feline communication but when it occurs indoors, not only is it unpleasant and unacceptable for us, if the cat is neutered it can also be a sign that he is stressed and insecure. If this is the case with your cat, it is important to see your vet to rule out any underlying medical reasons for the behaviour, and address possible stress issues that might need the help of a qualified and experienced behaviourist.

© Trudi Atkinson.

This appendix is published under a Creative Commons Attribution-NonCommercial-NoDerivatives 4.0 International Licence (http://creativecommons.org/licences/by-nc-nd/4.0/)

Appendix 12
Training Your Cat to Like the Cat Carrier

If your cat only associates the cat carrier with unpleasant car journeys that end up at the vets or cattery, it is not surprising if he becomes reluctant to go into the carrier. Using force to get him into the carrier can make matters even worse, often to the point where the cat becomes so fearful that he is likely to run away as soon as he sees the carrier. Therefore, it is better to teach him to accept the cat carrier by changing his perception of it, from being a dreaded device associated with being taken to frightening places, to being somewhere that is safe, comfortable, and secure.

If your cat already has a strong negative association with your current carrier it is probably best to purchase a new one with the following features:

- It should be made of a solid and easily cleaned material such as plastic and be large enough to accommodate your cat comfortably. It should have a front or side opening, and it is useful if it can also be opened from the top. Most importantly it should be possible to completely remove the top half of the carrier. (Fig A12.1)
- Ensure that the front or side opening door can be removed and replaced easily without needing to take the carrier apart. Also make sure that it is not hinged towards the centre which can partly block the opening.
- The clips separating the two halves should be sturdy and remain attached to the carrier. Small insubstantial clips can easily break, and clips that need to be removed can easily get lost.

Teaching your cat to accept the cat carrier

- If the carrier has been used before, even if it was a long time ago, wash and dry it well to remove any residual scent that might bring back bad memories for your cat.
- Remove the top half of the carrier and place some soft comfortable bedding in the bottom half (cats often prefer soft fleecy material, it can also help if it already contains the cat's own scent and/or your scent).
- Position the lower half of the carrier in a place that the cat already regards as safe, warm, and comfortable.
- Spraying synthetic facial pheromone into the carrier a few minutes before it is introduced to the cat may help the cat to perceive it as a safe place, and a few favourite treats may also help to encourage him towards it.
- Leave it in place so that the cat becomes accustomed to it and has the option to use it as a comfortable and secure bed.

Fig. A12.1. The best type of carrier to choose is one that can be separated into two halvesand has a front or side opening.

- It may take a few days or a few weeks for your cat to become fully relaxed and accustomed to the presence of bottom half of the carrier. Ideally, he should be using it regularly as a place to rest and relax. But if, after a few weeks he is still unwilling to go near the carrier: 1.) Ensure that it is situated somewhere safe and attractive for your cat. 2.) Try encouragements as described below.
- Once your cat is fully relaxed with the bottom half of the carrier, the top half can be replaced, but not the door.
- Even if previously accustomed to the bottom half, most cats are likely to be suspicious of the carrier once the top half has been attached, but given time some will be happy to go inside and may even continue to use it as a bed. For others, extra encouragement may be required, especially if the cat has had previous bad experiences with a cat carrier.

Encourage with food treats

Some cats can be encouraged into the carrier easily by throwing a few favourite treats inside the carrier or by poking a stick treat through the slats at the side or back of the carrier. Others however will need slow and gentle encouragement, as follows:

- Place a favourite food treat at a distance from the carrier where your cat feels relaxed. If he is reluctant or hesitant to take the treat move it further away from the carrier to where he will take the treat without any fear or hesitancy.
- Gradually place the treats closer to the carrier, but if the cat shows any fear or hesitance move the treat back to where it was previously. Take things slowly until the cat will happily go inside the carrier to take the treats.
- Using food treats frequently is recommended during training, but once training is complete treats should still be given, but intermittently rather than all the time as

this will help to maintain the learned behaviour. Also, there will be times when it is not possible to give your cat a treat.

Encourage with comfortable and safe bedding

This teaches the cat to regard the bedding within the carrier as safe and relaxing.

- Use a mat or 'flat' cat bed that will fit comfortably within the carrier. Ideally use something that your cat already chooses to sleep on.
- Initially place the bedding well away from the carrier and lure the cat towards the bedding by dropping food treats on to it.
- But the aim is not just to teach the cat to go to the bedding, we also need to teach him to settle and relax there.
- If the cat immediately lies down and relaxes on the bedding, this is ideal, and the cat should be rewarded with extra food treats. If he does not lie down and relax we may need to teach him to do so gradually.
- To start with the cat can be given additional treats just for standing on the bedding. But then treats should only be given for sitting and once he is sitting more often than standing, treats should then only be given for lying down.
- We also want to reward him for signs of relaxation such as washing and/or lying on his side. You may be able to help him relax by talking to him softly and stroking him, especially under his chin and on the sides of his face by the corner of his mouth.
- Once your cat is frequently lying down and relaxing on the bedding start to move it gradually closer to the carrier. *But never move the bedding while the cat is on it!*
- At each stage that you move the bedding closer to the carrier, lure the cat there with treats and encourage him to relax.
- Eventually the bedding can be put inside the carrier, but this also needs to be done gradually. Start by placing it just on the edge of the opening and then very slowly and gradually move it inside.

Encourage with toys

For some cats play may be the best way to reduce fear and to encourage them to go close to and enter the cat carrier.

- Use a favourite wand toy to play a game near to the carrier. As with food treats if your cat does not want to play, increase the distance from the carrier to where he is happy to play, and then slowly and gradually decrease the distance.
- If your cat is happy to play very close to the carrier it may then be possible to encourage him into the carrier by poking a suitable toy e.g. a feather through the slats in the carrier, or by dragging a string toy through the carrier via the slats at the back or side.
- The only problem with using play is that this increases general arousal and we really want the cat to be relaxed in and around the carrier. Therefore, once the cat's fear of the carrier has been reduced using toys, it is advisable to then use food treats and 'safe' bedding to encourage relaxation.

Shutting the door of the carrier

Once the cat is happy to go into the carrier without the door being attached it is then necessary to teach him to become accustomed to being shut in the carrier:

- Refit the door but leave it open and encourage the cat into the carrier as before.
- Keep the door open for a while (at least a few days for a nervous cat) to allow him to become accustomed to it before attempting to shut him in the carrier.
- Encourage him into the carrier using a treat and shut the door for no more than a second or two. Praise him, open the door, and give him another treat inside the carrier.
- Very gradually increase the time that the door is closed before opening it and giving him a treat.
- Treats can also be given to the cat while he is shut in the carrier by posting them through the closed door.
- If he shows any signs of distress or panic let him out straight away.

Travelling

Before taking your cat on a journey he needs to become accustomed to being moved from one place to another in the carrier:

- With the cat in the carrier move it slightly e.g. slide it a little way along the ground or lift it slowly and steadily up off the ground. Replace the carrier onto a firm surface then post a treat into the carrier for the cat.
- When you pick up the carrier with your cat inside hold it securely by the handle and underneath, rather than holding it by the top handle alone as any swinging motion can be uncomfortable for the cat.
- Start to get your cat accustomed to being transported in the carrier by carrying the cat around the house and garden, starting with very short distances i.e. just a few steps and then from one room to another. When you stop reward your cat by posting treats into the carrier. Also give him a treat when you let him out.
- The next stage is to place the carrier in the car and feed a few treats through the holes in the carrier. Do this first without the engine running, and then with the engine running.
- Secure the carrier well so that it does not move around during a journey. If there is enough room place the carrier on the floor rather than on a seat.
- Then take your cat for a few short journeys 'around the block' and back home again. Each time when you return home give your cat a special tasty meal or engage him in a favourite game.
- Covering the carrier with a lightweight towel or small sheet can block the cat's view of potentially frightening things, both during the journey and on arrival. But occasionally lift the cover to check that the cat is OK, and do not use anything that is too thick or too heavy that might restrict the air in the carrier.
- Drive steadily, trying as best you can to avoid bumps in the road or sudden breaking or acceleration.

- If the cat is accustomed to music or the radio being played at home, playing the same music or radio stations in the car may help to block out unfamiliar and frightening sounds for the cat.
- Never leave the cat unattended in the car, especially on a warm day.
- Never carry more than one cat at a time in a carrier.

This appendix was compiled with help from International Cat Care and I recommend that you watch the following videos:

Encouraging your cat to be happy in a cat carrier
Getting your cat used to travel
Putting your cat in a cat carrier

These videos can be accessed via the following link: https://icatcare.org/advice/cat-handling-videos or via the iCatCare YouTube page: https://m.youtube.com/user/iCatCare

Further Reading

Bradshaw J. and Ellis S. (2016) *The Trainable Cat. How to Make Life Happier for You and Your Cat.* Allen Lane, London.

© Trudi Atkinson.

This appendix is published under a Creative Commons Attribution-NonCommercial-NoDerivatives 4.0 International Licence (http://creativecommons.org/licences/by-nc-nd/4.0/)

Appendix 13
Medicating Your Cat

Mixing in Food

Liquids and powders can be easily mixed in with food. Small pills and tablets can also be hidden but larger tablets might need to be crushed. Some pills or tablets must be given whole, however, so always check with your vet and/or read the instructions you have been given before attempting to crush or break up a pill or tablet. Also check to ensure that the medication can be given with food.

Before attempting to give your cat his medication in food:

- Experiment to find a suitable food that your cat enjoys. All cats are different and not all cats will like the same food.
- If your cat has been prescribed a special diet or your vet has advised certain dietary restrictions, check with your vet to find out what food your cat can and cannot be given.
- It is often best to choose a food that is soft so that the medication can be easily mixed in with it. Food with a strong taste and smell will be better at disguising the taste and smell of the medication. Warming the food slightly to around body temperature can also increase the smell and make the food more attractive to the cat.
- Think about when and how often each day you will need to medicate your cat and start feeding him a small amount of this 'special' food at these times without any medication in it. This will teach him to expect a special treat at these times and reduce any suspicions he may have about a change in routine and being offered new food.
- Always feed the special treat before any other food so that your cat is hungry.
- The tablet can be crushed using a 'pill crusher' made especially for this purpose. You can see how to use one of these by watching the following video: https://www.youtube.com/watch?v=SWtnPyPQKaY. When you mix the medication into the food, ensure that there is sufficient food to disguise the medication but not so much that some of the food remains unmedicated and is eaten in preference to the medicated food.
- The medicated food can be given in a bowl or fed to the cat on the end of a spoon.
- Another option is to use food treats that are specially made for this purpose. These are usually semi-soft and designed so that a pill can be hidden inside. Similarly, something like cheese or other soft treats can be moulded around the pill or tablet. See https://www.youtube.com/watch?v=7Pqdcx0fQQU as to how to do this.

Physically Medicating the Cat

If a pill or tablet cannot be given with food and the only option is to physically medicate the cat, this is best if carried out by two people: one to hold the cat and the other to give the medication.

Advice for the person holding the cat

- Place the cat on a table or work surface that is covered with a towel or blanket.
- The cat should be facing away from you.
- Hold the cat into your body to prevent him from moving backwards and gently hold each leg just above the elbow, so that he is prevented from moving forward, or lifting his front legs.
- Using a towel:
  - Place the cat on top of a large towel.
  - Wrap the towel gently but firmly around the cat so that only his head is free.
  - Hold the cat as described above.

Advice for the person giving the pill or tablet

- Have the pill ready between your finger and thumb of your right hand (if you are right-handed).
- Using your other hand, place it over the cat's head so that your finger and thumb rest at the corners of the cat's mouth.
- Tilt the cat's head back slightly and by applying gentle pressure on the cat's lower incisors with a free finger of your right hand gently pull the lower jaw down and open the cat's mouth.
- Place or drop the pill as far back on the cat's tongue as you can.
- Lower the cat's head.
- Encourage the cat to swallow by syringing or spooning a little water into the side of his mouth. Do not squirt water directly down the back of the cat's throat. Giving the cat a small tasty food treat can also help to ensure that the medication has been swallowed.
- To see how to give a cat a pill or tablet watch the following video: https://www.youtube.com/watch?v=6_W0KTjElNs

Using a pill-giver or pill-popper

This is a syringe-like device that holds a pill or tablet and can be used to easily deposit it at the back of the cat's tongue, without needing to put your fingers into the cat's mouth. The following video demonstrates how to correctly use a pill-popper: https://www.youtube.com/watch?v=kkq_HKA7drA

Gelatine capsules

Empty gelatine capsules may be available via your vet practice that can be used to disguise the taste of bitter pills and can be very useful if multiple pills or tablets need to be given at the same time. These must always be washed down with water to prevent them from sticking in the cat's throat. Ensure that they are the correct size and do not make the medication too big for your cat to swallow easily.

Coating with butter

Coating with oil or butter may make a pill or tablet easier for the cat to swallow and may also help to disguise the taste.

Liquid Medication

Liquid medication or tablets crushed in water should be syringed or spooned into the side of the cat's mouth in small incremental amounts of around 0.5ml, as this is as much as a cat can comfortably swallow.

Applying a Spot-on Treatment

- It can help to prepare your cat well in advance by gently parting the hair at the back of his neck and then giving him a food treat while you are stroking him.
- Once the cat is unconcerned about having the hair on the back of his neck parted, dampen your finger and touch your wet finger on the cat's neck where the hair is parted and then give him a treat. This should allow him to become accustomed to the feeling of something wet on his neck.
- It can also help when you do apply the spot-on to warm it first by carrying it in your pocket for a while before applying it.
- Apply the treatment high up on the cat's neck so that he cannot lick it.
- If the cat needs to be restrained, do so as described previously.
- To see how to apply a spot-on treatment watch the following video: https://www.youtube.com/watch?v=xKlm-wGV2TU.
- Before you apply any spot-on treatment it is essential to make sure that it is suitable for cats, as some spot-on treatments for dogs can be highly toxic for cats.

If you continue to have difficulty in medicating your cat, speak to your vet or vet nurse/technician. They might be able to help you to medicate your cat, or your vet might be able to prescribe a different product containing the same or a similar medication in a form that is more palatable or easier to administer.

The video links in this appendix are from International Cat Care – https://www.icatcare.org. Other helpful cat-handling videos can be accessed via: https://icatcare.org/advice/cat-handling-videos or via the iCatCare YouTube page: https://m.youtube.com/user/iCatCare.

© Trudi Atkinson.

This appendix is published under a Creative Commons Attribution-NonCommercial-NoDerivatives 4.0 International Licence (http://creativecommons.org/licences/by-nc-nd/4.0/)

Appendix 14
Teaching your Cat or Kitten to Accept Veterinary Examination

Going to the vets and being physically examined can be an intrusive and unpleasant experience for many cats and kittens, unless it is something that the cat has become fully accustomed to and especially if it is an experience that has become associated with something enjoyable, such as being given tasty food treats.

- Start by preparing the food treats for your cat. These can be dry treats broken up into very small pieces or some form of soft food such as mashed up tuna, meat paste or yoghurt that can be given on a spoon. The food should be placed in a pot, preferably one with a lid, especially if your cat may attempt to steal the treats.
- Begin the training at a time when your cat is relaxed but not actually asleep.
- Concentrate on one area at a time at each training session.
- Keep your pot of treats (or soft food) nearby and reward with a treat immediately after each pretend examination.
- Start at a low level of handling and gradually increase so that your cat remains comfortable with your actions. Always ensure that your cat is fully comfortable with one stage of the training before progressing to the next.
- Never physically force your cat to comply, nor become angry or impatient with him. Stop immediately if your cat or kitten shows signs of distress or discomfort.
- Keep training sessions short – no more than a few seconds at a time – but try to repeat training sessions at least once a day.

Examining the Mouth

1. Start by gently stroking the cat's upper and lower lips a few times on one side, reward, then repeat on the other side. After doing this a few times, stroke both sides before rewarding.
2. Gently open the cat's lips very slightly, one side and then the other, then reward.
3. Do this daily – each time gradually lifting the lips and exposing the teeth a little bit more.
4. Also open the mouth – hold around the top of the cat's head gently, lift the upper lips slightly with your index finger and thumb, then place a finger or thumb from the other hand on the front of the lower jaw and pull this down to open the mouth. As before, start by opening the cat's mouth a very tiny amount and gradually increase at each training session.
5. To see how to do this watch the following video: https://m.youtube.com/watch?v=rHmnrULI4gU&feature=youtu.be.

Examining the Ears

1. Start by just stroking and gently rubbing the ears, then reward.
2. Progress to gently holding the tip of the cat's ear and lifting it slightly up and back to expose the ear canal, then reward.
3. Gradually increase this movement to where you can fold the ear flap back fully to expose and examine the ear canal. **Never put anything in your cat's ears unless advised to do so by your vet or vet nurse/technician.**
4. To see how to do this watch the following video: https://m.youtube.com/watch?v=4lGPATjs3no&feature=youtu.be.

Examining the Feet and Clipping the Nails

1. Start by just touching the foot, then reward.
2. Progress to gently holding the foot, then reward.
3. The next stage is to hold and gently rub the foot, gradually applying slightly more pressure so that you are eventually able to extend the claws. Be careful to take things slowly. It can help to give a food treat at the same time as doing this. Offering paste on the end of a spoon can often be a good idea as the cat will take time licking the paste.
4. Only when your cat is happy to have his foot held and claws extended should you progress to clipping his nails.
5. First allow your cat to become accustomed to the sight and feel of the clippers, by letting him investigate them and giving him a reward. Then just touch his feet with the clippers, followed by a reward.
6. When you clip his nails be careful to only clip the very ends of the claws to avoid hitting the blood vessel and nerve ending inside the nail. To begin with clip only a couple of claws or at most just the claws on one foot before giving a reward.
7. Gradually increase the number of nails you clip in between giving rewards. But always stop if your cat becomes restless or distressed.

To see how to accustom your cat to having his feet examined and his claws trimmed see: https://m.youtube.com/watch?v=V8SMinphtB4&feature=youtu.be, and: https://www.youtube.com/watch?v=U36rsW_WhUA.

The video links in this appendix are from International Cat Care – https://www.icatcare.org. Other helpful cat handling videos can be accessed via: https://icatcare.org/advice/cat-handling-videos or via the iCatCare YouTube page: https://m.youtube.com/user/iCatCare.

© Trudi Atkinson.

This appendix is published under a Creative Commons Attribution-NonCommercial-NoDerivatives 4.0 International Licence (http://creativecommons.org/licenses/by-nc-nd/4.0/)

Appendix 15
First Aid Advice for Common Feline Behaviour Problems

The following advice is designed to do no more than help manage the cat's behaviour problem in the short term and to help prevent the current problem from getting any worse. Specific advice aimed at resolving the problem cannot be given until a good understanding has been reached as to *why* the cat is behaving as he is, which can only be achieved through a combination of both behavioural and veterinary investigation.

General Advice for all Problems

Do not attempt to physically punish or reprimand the cat. This includes actions such as squirting the cat with water or shouting at the cat. A cat will not understand why you are angry and attempting to punish him is unlikely to be successful and more likely to make him frightened of you. If the attempt at punishment or reprimand causes the cat pain or fear the problem may escalate, or other more serious behaviour problems may develop (Fig. A15.1).

Behaviour problems can often be linked to medical disorders, so it is important that the cat is given a veterinary examination to rule out or treat any possible underlying physical cause or contributory factor. If problem behaviours persist or are severe it is important to seek professional help via your veterinary surgery from a suitably qualified feline behaviourist.

House Soiling

Is your cat going to the toilet (eliminating) in the wrong place or is he scent marking?

If your cat is going to the toilet in the wrong place:
- He will squat down to either urinate or defecate.
- You will find urine in puddles on horizontal surfaces, usually on the floor.

If he is eliminating, what are the first things to do?

- Make sure that the cat has access to at least one litter tray.
- Try placing a litter tray where the cat is currently going to the toilet. If the cat uses it, it can be moved to a more appropriate place later.
- Try providing additional litter trays – many cats prefer one area to urinate and a separate area to defecate so also make sure that there is sufficient distance between the litter trays.
- If you have recently changed to a different cat litter, go back to the old type.

Fig. A15.1. Do not attempt to punish the cat. It is unlikely to be successful and if the attempt at punishment causes the cat pain or fear, then the problem may become worse, or other behaviour problems may develop.

- Make sure that the litter trays are cleaned frequently and thoroughly:
    - Remove soiled and wet patches 2–3 times daily or as soon as they appear.
    - Clean and refresh the tray at least 1–2 times a week.
- Make sure litter trays are positioned well away from the cat's food, water, doors, windows or any place where the cat is likely to be disturbed.
- Make sure that the litter tray(s) are big enough to be comfortable for the cat, and that they contain enough litter for the cat to dig into and bury its waste.
- Clean soiled areas in the house with a 10% solution of biological detergent, rinse and then dab or spray lightly with surgical spirit (always test a small area first).
- Try to keep the cat away from areas where he normally soils; shut doors, block access with furniture or cover with a strong, impermeable and easy to clean covering (see Appendix 11).

If your cat is scent marking:

- Your cat will stand up and squirt urine backwards onto a vertical surface.
- You will find urine marks on walls and other vertical surfaces, plus sometimes a puddle underneath where it has run down the wall.

If he is scent marking (spraying), what are the first things to do?

- If the cat is unneutered, and is not to be used for breeding, arrange neutering as soon as possible. Urine marking is a normal behaviour and a part of sexual advertising in entire (uncastrated) male cats. Entire females may also scent mark this way when they are in season.
- Deny the cat access to the areas that are most often targeted unless this results in increased urine marking in other areas.
- Clean marked areas with a 10% solution of biological detergent or odour elimination product, rinse well and then dab or spray lightly with surgical spirit (always test a small area first).
- The use of a fraction 3 synthetic feline facial pheromone spray may help by altering the message conveyed in the scent mark. For this to be effective:
  1. Rinse the area really well after using the biological detergent or odour elimination product, or don't use any at all as these chemicals can also deactivate the pheromone.
  2. Spray the area well with the pheromone spray (4–5 sprays) every day for 3–4 weeks. Clean the area as before if the cat marks the same area again.
- Stress is most likely the underlying cause. Products designed to reduce or manage feline stress suggested by your vet or nurse/technician may help.
- Keep a diary to help identify possible sources of stress.

Aggression to People

Misdirected play

Not all biting and scratching is aggressive. When cats and kittens play, it is practice of predatory or defence skills, which means that teeth and claws are involved, so if play is directed towards us it can easily be mistaken for aggression.

How to tell if it is misdirected play:

- The cat is most likely to pounce on and bite moving hands and feet.
- There is unlikely to be any vocalization.
- The cat's body posture is likely to be forward and alert (Fig. A15.2).

Fig. A15.2. Typical 'play' body posture more likely to be seen if attacks on people are due to misdirected play.

What should you do?

- If the cat attacks you keep still. Movement can trigger further biting and scratching. When the cat releases his grip, withdraw your hand or foot very slowly.
- Try to re-direct the cat onto a moving toy.
- Do not encourage the cat or kitten to play with hands or feet.
- Increase play in general with toys that keep the hands well away from the part of the toy with which the cat is playing, such as wand toys (Fig. A15.3), motorized toys and 'bat and chase' toys.

How to tell if it is aggression:

- The cat is likely to become aggressive when approached or touched.
- The cat is likely to hiss, spit, growl or shriek.
- The cat's body posture is likely to be back and lowered, with ears flattened sideways or back (Fig. A15.4).

What should you do?

- Stop any interactions immediately, or do not attempt to approach your cat if he is showing signs of feeling frightened, agitated, or angry and likely to become aggressive, e.g. ears flattened backwards or to the sides; swishing tail; staring with dilated pupils; low growling and/or hissing; 'fluffed up' coat and/or tail (piloerection).
- Keep yourself and any other people or other animals safe by leaving the room but be careful not to block any potential escape routes for the cat.
- Try to keep calm, do not shout, lash out at the cat, or make any sudden movements which might cause the cat to attack.
- If the cat becomes aggressive during petting, keep petting times very short.
- Seek veterinary advice, and referral to a qualified behaviourist as soon as possible.
- In the meantime, try to avoid situations that might result in aggression, and wear 'tough' clothing for protection.

Fig. A15.3. Encouraging the cat to play with toys that do not involve human interaction, or keeping hands well away from the cat's teeth and claws can help to limit 'playful' predatory attacks on people. These toys can also be used to redirect the cat.

Fig. A15.4. A typical body posture of a fearful and defensively aggressive cat.

If someone is badly bitten or scratched:

- Wash the area well with soap and running water and seek medical attention immediately, especially if a child, elderly person, or an immune-suppressed individual has been bitten. Immediate medical attention is also essential if the individual experiences pain, swelling or redness around the wound, fever or headache after being bitten or scratched.

Fighting in a Multi-cat Household

Is it play or fighting?

If it is play:

- The cats are more likely to chase each other equally and swap roles.
- Claws remain retracted and bites are inhibited.
- There is no or very little vocalization.
- Following play the cats will want to remain near to each other, groom each other or rest very close or touching each other.

What should you do?

- Allow them to play and only intervene if it escalates into fighting.

If it is fighting:

- One cat is more likely to chase or 'pick on' the other.
- Claws are more likely to be extended.
- There is likely to be hissing, growling and/or 'shrieking'.

- They will generally try to avoid each other at other times or, if one cat is the victim, it will avoid or run away from the other cat.

What should you do?

- If the cats are fighting frequently and/or severely, separate them for 24–48 hours then attempt to reintroduce in the same way as you would introduce a new cat to the household (see Appendix 5). If fighting continues, separate them again until professional help is obtained.
- Feed the cats with a distance between them.
- Increase resource locations such as feeding areas, resting areas and litter trays (see Appendix 3).
- If fighting appears imminent place a physical barrier in between them.
- Be aware that cats sometimes re-direct aggression, so avoid handling or approaching cats that are highly aroused and aggressive.

Furniture Scratching

Cats need to scratch, both to help condition their claws by removing the outer 'dead' sheath, and as a way of both scent and visual marking. It is important therefore that cats are provided with suitable surfaces that they can scratch.

- Position a scratch post or pad near to the area that is currently being scratched.
- Pre-prepare the scratch post to encourage the cat to use it:
    - Transfer the scent from the currently scratched area by rubbing the post or pad over the area, or by rubbing a clean dry cloth over the area and then over the post or pad.
    - Pre-scratch the post with the tip of a screw.
    - Catnip or specially prepared feline synthetic interdigital pheromone with catnip can help to attract some cats to using the post or pad.
- Make the furniture less attractive for the cat to scratch (but do not do this until a viable alternative has been provided), by covering the area you do not want your cat to scratch with cling film, or a few short strips of double-sided sticky tape. If using tape test a small area first to ensure that it will not damage the furniture, and reduce the 'stickiness' so that it is unpleasant rather than uncomfortable for the cat when he touches it. A loose throw draped over a chair or sofa can also make it more difficult for a cat to get to and scratch the furniture.
- If scratching indoors is excessive it could be anxiety or stress related. The most common cause of feline stress is competition for resources with other cats. Following the advice contained in Appendix 3 may help if this is the case.

© Trudi Atkinson.

This appendix is published under a Creative Commons Attribution-NonCommercial-NoDerivatives 4.0 International Licence (http://creativecommons.org/licences/by-nc-nd/4.0/)

Appendix 16
Friend or Foe

Close social bonds can and do occur between cats, but not all cats that live together get along well with each other. Fighting is a clear sign that they are enemies but fighting can easily be misinterpreted as play, and even when there is no clear evidence of antagonism between them, this still does not indicate that they are friends.

Signs of a Friendly Relationship

- Grooming each other.
- Greeting each other by touching noses and with raised tails.
- Rubbing against each other.
- Wanting to be with each other.
- Sleeping curled up together.
- Playing together.

Signs of play

- Chasing each other equally and swapping roles.
- Claws remain retracted and bites are inhibited.
- No or very little vocalization.
- Following play the cats want to remain close to each other.

Signs of a Bad Relationship

- Trying to avoid each other:
  - Sleeping or resting in different rooms or in different areas of the same room.
  - Walking out of a room when the other cat walks in.
  - Jumping up onto furniture to avoid crossing paths.
  - Eating, sleeping or using a litter tray when the other cat is not around.
  - Spending more time outside when the other cat is inside.
- Guarding or blocking the other cat's access to important resources.
- Hissing, swiping, chasing or fighting.

Signs of fighting

- One cat is more likely to chase or 'pick on' the other.
- Claws are more likely to be extended and bites less inhibited.
- Hissing, growling and/or 'shrieking'.

Signs of a 'Tolerant' Relationship

There is another form of relationship between cats whereby they are not friends but they are able to tolerate each other. A human equivalent is rather like house sharing with someone who pays half of the rent but who is not a friend or family. The signs are:

- Few, if any, signs of a friendly relationship; unlikely to mutually groom or rub against each other.
- May occasionally greet each other in an amicable way.
- May share favourite resting areas but without touching each other.
- They do not make any great efforts to avoid each other but do not actively seek each other out either.

See 'Friend or Foe?', a video available from Cats Protection (https://m.youtube.com/watch?v=bPqreEUV5vM).

© Trudi Atkinson.

This appendix is published under a Creative Commons Attribution-NonCommercial-NoDerivatives 4.0 International Licence (http://creativecommons.org/licences/by-nc-nd/4.0/)

Appendix 17
Approaching, Stroking and Picking Up

How a cat is initially approached and handled, especially by someone they do not know well, can influence how the cat regards that person and what form and amount of physical interaction they are likely to feel happy with, although there will always be some degree of individual variation.

The Approach

It always best to allow the cat to feel in control of the situation and give him the option to approach and interact with you if he wants to. It is never a good idea to force interactions on him. However, you may need to decrease your distance from the cat and then encourage him towards you.

- If you need to decrease your distance from the cat do so calmly and slowly, avoiding fast or sudden movements.
- Keep some distance so that the cat has the option to approach you if he wants to.
- Avoid wearing strong perfume, as this may discourage the cat from approaching you, especially if you have scent on your hands.
- Avoid leaning over or reaching down to the cat. You will appear less threatening and the cat will be more likely to want to approach you if you get down to his level (Fig. A17.1).
- Avoid prolonged direct eye contact, especially if the cat is nervous. Position your body at an angle rather than directly facing the cat, so that you appear less threatening.
- Talk to the cat using a soft, calm and measured voice.
- Avoid loud, abrupt or shhh sounds that the cat may perceived as threatening (such noises may sound similar to another cat hissing).
- Use a 'slow eye blink', as follows:
  - When the cat looks towards you, close your eyes slowly and keep them partially closed for a couple of seconds.
- Offer the cat the back of your hand to sniff, before attempting to touch the cat.

If the cat moves away or shows no sign of wanting to be stroked, i.e. by pushing his head towards your hand or rubbing against you, then it is best to leave him alone. If he does show signs of wanting to be stroked do so as follows:

- Stroke gently under his chin and along the side of his face.
- Do this for no more than a few seconds then stop, but keep your hand close to his face, giving him the option to continue with the interaction (by pushing his head towards your hand) or not. Leave him alone if he moves away or does not push his head towards you.

Fig. A17.1. Avoid leaning over or reaching down to the cat. Try to approach on the same level as the cat and offer the cat your hand to sniff.

- Speak to the cat while stroking, keeping your voice calm, soft and relaxed.
- As the cat becomes more relaxed, try stroking other areas of his body, although always start by stroking around the face and under the chin. Cats vary as to where they like, or will accept being touched or stroked, so this can be a matter of trial and error. However, many cats do not like to be touched, especially by a stranger, on the feet or belly.

A cat may perform what is known as a 'social roll' whereby the cat rolls on his back while stretching and opening the claws. This should not be interpreted as a request by the cat to have his 'tummy rubbed', as the cat may want to play and not be fussed. Also, trying to touch the cat's belly could be considered threatening by some cats and provoke an aggressive defensive reaction.

Immediately stop stroking and move away if the cat shows any of the following behaviours:

- Dilated pupils.
- Swishing tail.
- Swiping at you or pushing your hand away.
- Biting.
- Twitching skin.
- Hissing or growling.
- Backing off or moving away.
- Ears flattened to the side or back.

Picking the Cat Up

Avoid picking the cat up unless it is absolutely necessary to do so, or if you are certain that the cat wants to be picked up (although it can be easy to misinterpret such signals unless you know the cat well).

If you do need to pick the cat up:

- **Never** pick the cat up by the scruff of the neck. This can be painful and frightening for the cat and may produce a defensive and aggressive reaction.

- Ensure that the cat is well supported underneath; do not 'dangle' the cat by only holding it around the chest.
- If the cat shows any sign of wanting to be released, do so immediately; do not restrain the cat.
- It is better to allow and encourage the cat to come to you and to jump up onto your lap, rather than attempting to force the cat into any form of physical interaction or restraint.

I recommend that you also watch the following videos produced by International Cat Care:

Approaching a cat
How to touch and stroke a cat
Things to avoid when handling a cat

These videos can be accessed via the following link: https://icatcare.org/advice/cat-handling-videos or via the iCatCare YouTube page: https://m.youtube.com/user/iCatCar

© Trudi Atkinson.

This appendix is published under a Creative Commons Attribution-NonCommercial-NoDerivatives 4.0 International Licence (http://creativecommons.org/licences/by-nc-nd/4.0/)

Appendix 18
Recognizing Stress

It is not easy to spot if a cat is stressed because what is stressful to one cat may not be in the least bit stressful to another, and every cat will react differently. Also, because demonstrating any weakness could make them vulnerable to attack from potential predators, or even other cats, they can be very good at hiding their feelings, especially when they feel threatened.

However, any of the following signs could indicate that a cat maybe suffering from stress.

Physiological Signs of Stress

These signs are more likely to be seen as a result of acute stress:

- Panting.
- Salivation.
- Dilated pupils.
- Sweaty paws.
- Loss of bladder control.
- Diarrhoea.
- Constipation.
- Loss of appetite.

Behavioural Signs of Stress

- Increased vigilance.
- Appearing tense, only able to relax for short periods.
- Lowered head and body posture, especially when the cat is fearful. The head may be positioned lower than the body.
- Ears flattened sideways.
- Tail held close to body.
- Poor appetite.
- A decrease in grooming, resulting in a matted, uncared for coat, or 'excessive grooming' causing regular regurgitation of hair balls, and if excessive, hair thinning and bald patches.
- Withdrawing – not wanting to interact, play or be fussed.
- Continually or frequently attempting to hide or escape.
- Sham sleeping. If a cat can't escape the thing that is making him feel stressed, he will often pretend to sleep and hope that whatever or whoever it is will just go away.
- Indoor urine marking (spraying) is a common sign of stress in neutered house cats.

© Trudi Atkinson.

This appendix is published under a Creative Commons Attribution-NonCommercial-NoDerivatives 4.0 International Licence (http://creativecommons.org/licences/by-nc-nd/4.0/)

Glossary

Affiliative behaviour: Friendly behaviours that increase or maintain a good relationship between individuals, for example allogrooming and allorubbing.
Allo: From the Greek *allos* meaning 'other' (e.g. allogrooming is grooming another as opposed to self-grooming).
Catecholamines: A group of organic compounds that function as hormones, neurotransmitters, or both. They include dopamine, adrenaline (epinephrine) and noradrenaline (norepinephrine).
Conspecific: A member of the same species.
Coping strategy: A means by which an individual may attempt to lessen or tolerate a negative emotional state.
Displacement activity: A behaviour that is performed unnecessarily or out of context, e.g. scratching or grooming, typically at a time of mild to moderate emotional conflict.
DNA: Deoxyribonucleic acid. The main component of chromosomes and the carrier of genetic information within a cell.
Dyad: Consisting of two parts; a group of two individuals.
Emotional conflict: The presence of conflicting emotions resulting in distress, e.g. fear or frustration connected with an attached or trusted individual.
Endocrine: Concerning glands that secrete hormones directly into the bloodstream.
Epithelium: Protective covering of external and internal surfaces, for example skin and the lining of internal vessels and cavities.
Felidae: The cat family.
Ganglion *(pl.* ganglia): A collection of nerve cells and fibres combined in a single structure.
Genetic: Concerning genes or heredity.
Glycogen: The form in which glucose is stored within the body.
Homeostasis: Maintenance of a stable internal environment.
Hormone: A chemical substance secreted by one organ and carried in the bloodstream to another organ on which it has a specific effect.
Hydrolysis: Chemical breakdown that requires water.
Idiopathic: Of unknown cause.
Inherent: A physical or behavioural characteristic that exists at birth, as opposed to one that is **'acquired'**, i.e. develops due to influences after birth.
Limbic system: An area of the brain situated deep within the brain cortex involved with the processing of emotions. The primary structures of the limbic system include the amygdala, hippocampus, basal ganglia and cingulate gyrus.
Metabolism: Organic processes that convert food into energy.

Neoteny: The retention of juvenile characteristics into adulthood. **Neotenic** – related to neoteny.
Neuropathic pain: Pain that is a result of damaged or dysfunctional nerve fibres.
Oestrus: Period of sexual receptivity and fertility in female mammals.
Pathological: Causing or resulting from disease.
Photoreceptor: A specialised nerve cell or group of cells sensitive and responsive to light.
Physiological: Normal functioning of living organisms.
Piloerection: Contraction of muscles in the skin that raise the hair follicles so that the hair appears to stand up on end, often a response to a threat, making the animal appear bigger than it really is.
Psychogenic: Originating in the mind.
Resource: Anything that is necessary or desired by the individual, e.g. food, water, resting places, elimination areas, affection.
Selective breeding (aka **artificial selection**): Breeding of animals or plants by humans to select, enhance or maintain aspects of physical appearance or behavioural temperament.
Stimulus (*pl.* **stimuli**): Something that can cause or evoke a physiological response. The senses are responsive to the external stimuli sight, sound, smell and touch.
Synthesis: The production of a more complex structure from simple components.
Taxonomy: Classification of organisms based on similarities of structure or origin (adjective: **taxonomic**).

List of Useful Websites

Animal and Behaviour Training Council: www.abtcouncil.org.uk
ASAB Register of Certified Practitioners: www.asab.org/ccab-register
Association of Pet Behaviour Counsellors: www.apbc.org.uk
Cat Professional (specialist feline veterinary care and advice): https://www.vetprofessionals.com/site/cat-professional
Cats Protection: https://www.cats.org.uk
Celia Haddon (online advice for behaviour problems): www.catexpert.co.uk
Food Puzzles for Cats: http://foodpuzzlesforcats.com
International Association of Animal Behaviour Consultants: https://iaabc.org
International Cat Care: https://icatcare.org
Protectapet (cat proof fencing): https://protectapet.com

Recommended Reading List

Bradshaw J. (2013) *Cat Sense: The Feline Enigma Revealed*. Penguin Books, London.
Bradshaw, J. and Ellis, S. (2016) *The Trainable Cat: How to Make Life Happier for You and Your Cat*. Penguin Books, London.
Bradshaw, J.W.S., Casey, R.A. and Brown, S.L. (2012) *The Behaviour of the Domestic Cat*, 2nd edn. CAB International, Wallingford, UK.
Ellis, S. and Sparkes, A. (2016) *ISFM Guide to Feline Stress and Health: Managing Negative Emotions to Improve Feline Health and Wellbeing*. International Cat Care, Tisbury, Wiltshire, UK.
Rodan, I. and Heath, S. (2016) *Feline Behavioural Health and Welfare*. Saunders, Elsevier, St Louis, Missouri, USA.

Index

Note: bold page numbers indicate figures; italic page numbers indicate tables.

AAFP (American Association of Feline Practitioners) 151
ABTC (Animal Behaviour and Training Council) 169, **169**
ACTH (adrenocorticotropic hormone) 83
acupuncture 176, 177
adoption 188–189
adrenal glands 83
adrenaline 83, *172*
African wildcat (*Felix silvestris lybica*) 3, 5, **5**, 6, 22, 42, 49, 62
age-related issues 79, 81–82, 89
aggression 31, 33, 58, 145
    and breeding 121, 124
    defensive 150, 247–248
    and epigenetics 67
    in hand-reared cats 66
    human-directed 11, 91, 150, 247–249, **247**, **248**
    maternal 59, 65, 72, 121, 124, 129, **129**
    misinterpreted as play 45, 46, 48, 75, 200–201, 251–252
    in multi-cat households 45
    and pain 80
    self-directed 89
    and stress 90–91
    and visits to vet's 151, 152, 153, 155, 158
    and visual signals 30, **31**, 32
    and vocalizations 23, 26
    see also fighting
air-fresheners 155, 159
air-righting reflex 19–20, **20**
alleochemicals 35
allogrooming 33, 34–35, 45
allorubbing 33–34, **34**, 45, 173
ambush attacks 48, 184, 224, 231
American Association of Feline Practitioners (AAFP) 151
amino acids 51–52, 58, *175*
Animal Behaviour and Training Council (ABTC) 169, **169**
anoestrus phase 60, 62
anogenital reflex 69
anthropomorphism 11

anxiety 67, 81, 82, 90
    in kittens 26
    medications for *171*, *175*
    separation 50, 51
    and spraying 38–39, 91
Arabian wildcat *see* African wildcat
aromatherapy 176–177
Artemis (Greek deity) 8
arthritis 79
associative learning 100–112
    and classical conditioning 100–101, 111
    and counterconditioning 111
    and food treats 101, 104, 108, 111, 114
    and modifying unwanted behaviour *see* unwanted behaviour, modifying
    and operant conditioning/instrumental learning *see* operant conditioning
    and redirecting behaviours/alternative targets 112
    and shaping behaviour/successive approximation *see* shaping behaviour
attachment 10–11
attitudes towards cats 10–11
    historical 7–8
autonomic nervous system 83
avoidance behaviours 47, 94

babies/children 149, 227–229
Bach flower remedies 174–176, 177
balancing skills 69
Bastet (Egyptian deity) 7, 8
bedding 182, **182**, 187
behavioural first aid 165–168, 245–250
    advice to give/not to give in 167
    and cat's needs 166–167
    and client expectations 166
    and client's needs 166
    and time/resource limitations 165–166
behavioural pharmacology 169–170, *171–172*, **173**
    see also pheromonatherapy

behavioural problems/treatment 165–178
    and CAM *see* complementary and alternative medicine
    and initial consultation *see* behavioural first aid
    online advice for 167
    and owner's behaviour 178
    and placebo effect 178
    and referrals 168–169
    and regulations/bodies 169, **169**
    time required for full consultation 166, 168
big cats (*Panthera*) 3, *4*
black-footed cat (*Felix nigripes*) 3, *4*
bladder inflammation 84
blinking 28, **29**
blocking behaviour 47–48, **48**, 211
body temperature, in kittens 68
brachycephalic cats 9
breeding 119–130
    and care of stud cat 120
    and education of new owners 129–130
    and habituation 126
    and hand-rearing 124–125, **125**
    and handling kittens *see* handling kittens
    and hygiene 128
    and kittens' early experience 126–128
    and maternal aggression 129, **129**
    online advice for 119, 124
    and parturition *see* parturition
    pre-weaning period 64–65, **64**, 123–124
    and pregnancy *see* pregnancy
    and prevention of behaviour problems 119
    selection of queen/stud for 119–120
    selective, and feline behaviour 8–9, **9**
    and socialization *see* socialization
    and vaccinations 120, 128, 129
    weaning *see* weaning
Britain (UK)
    cat breed registries in 8
    cat ownership in 10
Burmese 89

CAM *see* complementary and alternative medicine
CAP (cat-appeasing pheromone) 173–174
cardiovascular system 90
carpel hairs 18
castration 44
cat breed registries 8
cat cafés 190–194, **191**
    alternative to 194
    control of customers in 193–194
    design of/equipment for 193
    and qualified/expert support 192
    sourcing cats for 192
    stability of groups in 192–193
    staff suitability for 193
    stress/disturbance in 191–192
cat carers, advice for 180–194
cat carriers 101, 111, 139, 152, 153, 162, **236**
    removing cat from 156–157
    training cats to like 235–239
Cat Fanciers Association (CFA) 8, 9
cat flaps 106, 144, 149, 223–225
    training cat to use 225
    types of 224
cat fosterers 187–188, 194
cat owners, advice for 141–151
    avoiding aggression 150
    and babies/children 149, 227–229, **228**
    and behavioural problems *see* behavioural consultation
    and cat flaps 106, 144, 149
    and cat health 151
    and dogs 137, **137**, 149
    and house-training *see* house-training
    and indoor lifestyle/outdoor access 141–142
    and leads/harnesses 145
    managing stress 141
    and multi-cat households 146
    and neighbourhood cats 146–148, **146**
    and owners' responsibilities 134
    and perceived threats from outside 148–149, **148**
    prospective *see* prospective owners, advice for
    and semi-confinement *see* containment systems
    see also cat–human relationship; veterinary professionals–client interaction
cat ownership, reasons for 10–11, 131–133
    health benefits 11, 132
cat shelters *see* catteries/shelters
cat sitters 180, 189–190
cat species *4, 5*
cat trees 132, 148, 195, **196**, 228
cat-appeasing pheromone (CAP) 173–174
cat–human relationship 49–51
    approaching/picking up 253–255
    and behavioural problems 166
    and mother–kitten relationship 49–50, 72
catecholamines 83, 85, 172, 257
catnip (*Nepeta cataria*) 175, 177, 204, 250
catteries/shelters 88, 89, 114, 131, 170, 180, 180–187
    admissions process 185–186
    bedding/sleeping areas in 182, **182**, 187
    and cat cafés/cat-themed establishments 192–193, 194
    cleaning regimes in 187
    communal housing in 183–185
    consistency/routine in 186
    and dogs 186
    housing construction/design/furniture in 181–182

housing location in 180–181
human contact in 185
introducing new cats to 184–186
minimizing stress/conflict in 184
private areas/hiding places in 181, **181**, 184
and stress 86, 182, 183, 184, 186, 189
CDS *see* cognitive dysfunction syndrome
central nervous system (CNS) 170, *171*, 177
and disease 81, 92
CFA (Cat Fanciers Association) 8, 9
chattering/chittering 22, 25
children 137, 149, 227–229, **228**
cilia 19
classical conditioning 100–101, 111
and counterconditioning 111
claws 19, 31, 35, 54
digging into owner's leg 109
examining/clipping 243
clickers 106, 107
Code of Practice for Welfare of Cats (2006) 134
cognitive dysfunction syndrome (CDS) 81, 82, 90, 92, *172*
and learning/training 115
communication 22–39
auditory *see* vocalizations
methods, advantages/disadvantages of 23
scent signals *see* olfactory signals
tactile *see* tactile communication
visual *see* visual signalling
complementary and alternative medicine (CAM) 174–178
acupuncture 176, 177
aromatherapy 176–177
Bach flower remedies 174–176, 177
efficacy of 177–178
homeopathy 174, 177
nutraceuticals/herbal supplements 175
TTouch 177
zoopharmacognosy 176, 177
compulsive behaviours 92
conditioned reinforcement
see secondary/conditioned reinforcement
conditioned response/stimulus (CR/CS) 101
conflict avoidance 27, 45, 49
constipation 88, 256
containment systems 142–145
cat-proof fencing 120, 142, **143**, 147, **147**
electric 142–144
purpose-built enclosures 144–145, **144**
context-specific learning 114
core territory 39, 44, 45, 49, 162, 211
corticotropin releasing hormone (CRH) 83
cortisol 89, 93
counterconditioning 111
CR/CS (conditioned response/stimulus) 101
cross-eyed Siamese 16

cysteine 51–52, 58
cystitis 84–85, 115

defecation
and disease 81
house-soiling 11, 81, 82, 91
and pain 80
and stress 93
see also litter trays
deferential signalling 27, 31, 49
degenerative joint disease 79, 81
dermatitis 86, 87
dermatological conditions 86–87
desensitization 111
diabetes 88–89, *172*
Diana (Roman deity) 8
diarrhoea 88, 256
diet 51–52, 58, 151, 177
and cats as obligatory carnivores 52, 53, 177
and nutraceuticals/herbal supplements 175
for pregnant queens 121
discriminative stimulus 113
disease 81, 134, 135, 142, 146
and behaviour *see* health and behaviour
and communal living 43, 183
and stress 84–90, 153, 158
displacement activities 25, 86, 257
distress *see* stress
dogs 11, 23–24, 26–27, 51, 90, 131, 149, 186
and aromatherapy 177
introducing 217–222, **218**
and learning/training 98, 100–101, 106, 220
and veterinary surgeries 153, 155
domestication 5–7, **6**, 42, 49
dominance hierarchies 48–49
dopamine 170, *172*, 177

early socialization *see* socialization
ears
signalling using 27, **28**, 93, 158, 256
vestibular system 19
veterinary examination of 244
see also hearing
Egypt, ancient 7
electric containment systems 142–144
endocrine system 88–89
enteritis 85
environmental enrichment 195–198, **198**
cat trees 132, 148, 195, **196**, 228
puzzle feeders/foraging games 125, 182, 195–199
for stud cats 120
walkways 193, 195, **197**
see also hiding spaces, provision of; shelving/elevated areas

Index 267

epigenetics 67
epinephrine 83
ethmoturbinals 17
European wildcat (*Felix silvestris silvestris*) 3, 5
eustress 82
evolution/origin of cats 3–11
    and cat species 3, *4*, 5
    and current attitudes towards cats 10–11
    domestication process 5–7, **6**
    and historical attitudes towards cats 7–8
    and hybrid breeds 9–10, *10*
    pedigree breeds 8–9
extinction/extinction burst 110, 111
eye contact 28, 30
    human–cat 157, 161, 253
eyes
    retina 14
    rods/cones 14, 15–16
    signalling with 27–28, **28**
    see also pupils; vision

facial 'expressions'/movements 27–29, **28–29**
farm cats 24, 34, 39
FCoV (feline coronavirus) 85
FCV (feline calicivirus) 86
fear 27, 28, **28**, 29, 30, **31**, 79, 141, **249**, 256
    and kittens 68, 70, 71, 72
    and learning 101, 109, 110, 111, 114
    and mental health 90
    and pain 79, 80
    and sensitization 100
    signalling 27, 28, **28**
    of vet's 151, 152–153, 154, 155, 159, 160
feeding behaviour 51–52
    and feeding patterns 52
    in feral cats 43, 52
    in multi-cat households 46–47, **47**, 205
    and new food 52
    and stress 88, 256
    and taste receptors 52
    and temperature of food 52
    and water dishes 121, 184, 206, 214
    see also diet; puzzle feeders/foraging games
feet/foot pads 18–19
    and olfactory communication 37
    tapping/pawing signals 35
Felidae 3, 4, 257
Felifriend 173
feline calicivirus (FCV) 86
feline coronavirus (FCoV) 85
feline herpesvirus (FHV/FeHV-1) 85–86
feline hyperaesthesia syndrome 89–90
feline infectious peritonitis (FIP) 85, 92
feline interdigital semiochemical (FIS) 174
feline lower urinary tract disease (FLUTD) 84, 86
feline oral facial pain syndrome (FOPS) 89, 90

felinine 58
Feliscratch 174
Feliway Classic 170–173
Feliway Friends/Multi-Cat 173
Felix chaus 4, 10
Felix silvestris lybica see African wildcat
Felix silvestris silvestris 3, 5
fencing, cat-proof 120, 142, 147, **147**
feral cats 23, 24, 33, 34, 39, 192
    dominance hierarchies in 49
    effect on wildlife of 54–55
    feeding behaviour in 43, 52
    home ranges of 44
    long-sightedness in 69
    mature males 43, 44
    and parturition 63
    resource sharing of 43
    size of colonies 43
    social behaviour of 42–44
FHV/FeHV-1 (feline herpesvirus) 85–86
FIC (feline idiopathic cystitis) 84–85
'fight or flight' response 83, 90, 93
fighting 49, 58, 91, 146
    misinterpreted as play 45, 46, 48, 75, 200–201, 251–252
    and multi-cat households 45, 249–250
FIP (feline infectious peritonitis) 85, 92
fireworks 100, 126
FIS (feline interdigital semiochemical) 174
flehmen response 17, **18**, 38, 68
FLUTD (feline lower urinary tract disease) 84, 86
follicle stimulating hormone (FSH) 61
food *see* feeding behaviour
food puzzles/foraging 125, 182, 195–199
food treats 101, 104, 108, 111, 114, 157, 226
FOPS (feline oral facial pain syndrome) 89, 90
foster care 187–188, 194
frustration/irritation 90, 91, **159**, 182
    signalling 27, **28**, 29

GABA (γ-aminobutyric acid) 170, *171*, 175
gastrointestinal system 87–88, 91, 92
GCCF (Governing Council of the Cat Fancy) 8, 9, 119, 138
genal whiskers 18
Geoffroy's cat (*Leopardus geoffroyi*) 4, *10*
gestation period *61*, 62
glucocorticoids/glucocorticosteroids 83, 85, 88, 93
glutamate 170, *171*, 175
glycogen 83, 257
Governing Council of the Cat Fancy (GCCF) 8, 9, 119, 138
grieving cats 133, 215–216
grooming 33, 34, 35, 45, 80
    excessive, and stress 86, **87**
growling 25, 46

habituation 71–72, 99, 111, 126, 136, 187
hairballs 86, 88
hand-rearing 124–125, **125**
handling cats 138, 152, 154, 155–157
    picking up 254–255
handling kittens 65, 69, 71, 127, **128**, 137
    frequency of 128
head-butting 35
health and behaviour 79–94, 141, 151, 166
    and learning/training 115
    and old age 81–82
    and pain 79–80
health care 121, **121**, 134, 151
hearing 16–17, 54
    in kittens 68
    and pinna 17
hepatic lipidosis 88
herbal supplements *175*
hiding 122, 150
    and stress 94, 114, 158, 256
hiding spaces, provision of 122, 139, 145, 159, 160–161, 181, **181**, 184, 193, 195
hissing 25, 26, 46, 201, 213, 214, 254
hissing sounds when talking to cat,
    avoiding 107, 157, 253
homeopathy 174, 177
hormones 257
    cortisol 93
    CRH/ACTH 83
    follicle stimulating (FSH) *61*
    luteinizing (LH) 60, *61*
    oxytocin 50–51
    stress 67, 84, 85, 120–121
    thyroid 89
hospitalization 158–162
    and cleaning regime 160–161
    housing/caging for 158–159
    nursing care for 160
    removing from cage 161
    returning home following 161–162
    stress in 158, **159**, 160
house mouse (*Mus musculus*) 6
house-soiling 11, 81, 82, 91
house-training 149, 230–234
    and accidents 232–233
    and outdoor toilets 232
    and scent-marking 233–234
HPA (hypothalamus–pituitary–adrenal) axis 83
hunting 53–55
    and body design/faculties 54
    and body position 53, **54**
    effect on wildlife of 54–55
    and kittens 65–66, **65**, 74, 125
    and mystacial vibrissae 18
    and tail signals 29
hybrid breeds 9–10, *10*
hyperaesthesia 89–90

hyperthyroidism 81, 89
hypothalamus 83, 176

IBS (irritable bowel syndrome) 87
immune system 85–86
in-house treatment 167–168
    and RCVS code of conduct 168
independence of cats 10, 51
indoor only housed cats 141, 142, 192
induced ovulation 60
infanticide 59, 63
instrumental learning *see* operant conditioning
International Cat Care 119, 162
interoestrus 60, *61*
irritable bowel syndrome (IBS) 87
irritation *see* frustration/irritation, signalling
ISFM (International Society of Feline
    Medicine) 151, 164
Isis (Egyptian deity) 8

Jacobson's/vomeronasal organ 17, **18**, 35, 68
jungle cat (*Felix chaus*) 4, *10*

kitten play 72–74
    locomotor 72, 74
    object 72, 74
    social 72–73, 73, **73**
    and socialization 71
kittens 64–74, 187
    advice for prospective owners of
        136–138, **137**
    and allorubbing 34
    as altricial species 67, **68**
    behavioural development in 69–72
    birth of *see* parturition
    and breeders 119
    feeding behaviour of 52, 121
    frustration experienced by 66, 125
    and habituation 71–72
    hand-reared 66, 124–125, **125**
    influence of mother/littermates on 72
    inherited personality traits in 66–67
    introducing to other cats 210, 211
    and kneading 35
    learning ability/sensitive period in 69, 70,
        72, 126–128
    and olfaction 63
    postnatal physical development 67–69
    pre-birth 66–67
    pre-weaning period 64–65, **64**,
        123–124, 136
    and socialization *see* socialization
    toileting of 69
    vocalizations of 26

kittens (*continued*)
    and weaning 52, 65–66, **65**, 121
    see also mother–kitten relationship
kneading 35, 49–50

Lamellar/Pacinian corpuscles 19
laser pens 202
lavender oil 177
leads/harnesses 145
learning theory 98, 109
learning/training 98–115
    associative *see* associative learning
    benefits of 98, **99**
    context-specific 114
    and discriminative stimulus 113
    factors influencing 112–115
    and habituation 71–72, 99, 111
    and health/cognitive abilities 115
    and learned helplessness 114
    and motivation 112–113
    and overshadowing 113–114
    and sensitization 100
    and sound/clickers 105–106, 107, 113, 114, 226
    and superstitious behaviour 114
    teaching to come to owner 226
    and 'training toolbox' 113
leopard cat (*Prionailurus bengalensis*) 4, 10
Leopardus wiedii 4, 25
Leptailurus serval 4, 10
licking 35, 89
limbic system 83, 176, 257
lion (*Panthera leo*) 3, 4
litter trays 82, 115, 132, 134, 139, 145, 149, 187
    and adoption/cat sitting 189, 190
    and breeding queens 119, 122, 123
    and cat cafés 193
    and cat flaps 223
    in communal enclosures 184
    and hospitalized cats 158, 159
    learned aversion to 80
    in multi-cat households 45, 47, 48, 146, 206
    and pregnant queens 122
    and shelters/catteries 184, 185, 187
    see also house-training
locomotor play 72, 74
long-haired breeds 8
luring **106**, 106, 108
luteinizing hormone (LH) 60, *61*

margay (*Leopardus wiedii*) 4, 25
mastitis 64
mating 59
medication 169–170, *171–172*
    administering 164, 170, 240–242
    see also pheramonatherapy

mental/emotional health 90
meowing 22, 23–24
    silent 24
    specific to owners 24
methionine 51–52, 58
metoestrus phase 60, *61–62*
Middle Ages 7–8
Middle East 5–6, **6**
monks/monasteries 7–8
mood changes 80, 81
mother–kitten relationship
    aggression in 72, 121
    and cat–human relationship 49–50
    and feeding behaviour 52
    pre-weaning period 64–65, **64**, 123–124
    and social learning 72
    and vocalizations 26
    and weaning *see* weaning
motor development 69
mouth ulcers 89
multi-cat households 44–49, 89, 146, 162, 183
    dominance hierarchies in 48–49
    fighting/aggression in 45, 46, 47–48, 249–250
    food/water in 46–47, **47**, 205–206
    litter trays in 45, 47, 48, 132, 146, 206
    misleading appearance of social bonding in 46–47
    and new introductions 133, 210–214
    and pheromonatherapy 173
    reducing resource competition in 205–206, 214
    resting places in 45, 47, 206, 214
    social bonding in 45–46
    space requirements for 132
Munchkin **9**
Mus musculus 6
mutations/deformities 8–9, **9**
    cross-eyed Siamese 16
mystacial vibrissae 18, 28–29

nasal turbinate bones 17
nasopalatine ducts 17
neoplasia 84
Nepeta cataria (Catnip) *175*, 177, 204, 250
neurological conditions 86, 89–90
neurotransmitters 170, *171*, **173**
neutering/neutered cats 44, 135, 145–146, 207–209
    and aggression 45, 145, 208
    and scratching 37
    and spraying 38, 58, 91, 145, 208
    when to 208
night vision 14–15
noradrenaline/norepinephrine 83, 170, *171*, 176
North African wildcat *see* African wildcat
nutraceuticals/herbal supplements 175

obesity 88, 89
object play 72, 74
oestrus cycle 60, *61–62*
olfaction 17, **18**, 52, 161
    and aromatherapy 176–177
    and kittens 63, 68
    and scent swapping 213
    see also pheromonatherapy; pheromones
olfactory signals 22, 35–39, 161
    advantages/disadvantages of 23
    alleochemical 35
    and allorubbing 34
    faeces marking 39
    pheromone *see* pheromones
    semiochemical 35, 174
    and skin glands 36–37, **37**
    spraying *see* urine spraying
    territory marking 37, 39, 91
operant conditioning 101–107
    and positive/negative reinforcement 102
    and rewards/reinforcers 102–103, 104
    and secondary/conditioned reinforcement *see* secondary/conditioned reinforcement
    and self-rewarding behaviour 103
    and unintentional reinforcement 103
origin of domestic cats *see* evolution/origin of cats
osteoarthritis 79
osteochondrodysplasia 9
otolith organs 19
overeating 88
overshadowing 113–114
owner reinforcement 24
oxytocin 50–51, 176

Pacinian/Lamellar corpuscles 19
pain 79–80, 81, **81**, 86, 89, 115
    problems discerning 79
Panthera 3, 4
panting 62, 93, 256
parasites, 86, 138
    treatments for 151
parturition 123
    imminent, signs of 62–63
Pavlov's conditioned reflexes 100–101
pedigree cat breeds 8–9
penile spines 59
Persian **9**
personality traits, inherited 66–67
pheromonatherapy 170–174
    cat-appeasing pheromone (CAP) 173–174
    feline interdigital semiochemical (FIS) 174
    synthetic facial pheromones 170–173
pheromones 35–36, 59, 170
    as social odours 35

photoreceptors 14, 15–16
pica 92
pied tamarin (*Saguinus bicolor*) 25
piloerection 29, **31**, 93
pinna 17, 27, 54
placebo effect 178
plantar pad 37, 174
play 28, 29, 31, 45, 72–75, 200–204
    aggression misinterpreted as 45, 46, 48, 75, 200–201, 251–252
    avoiding attacks on humans in 150
    cat–human 50
    ending 204
    and hunting skills 66
    and kittens *see* kitten play
    locomotor 72, 74
    object 71, 72, 74, 201–202
    signs of cat's desire to 202, **203**
    social 72–73, *73*, **73**, 75, 200
    and training 104, 105, 112, 113
    see also puzzle feeders/foraging games; toys
pre-weaning period 64–65, **64**, 123–124, 136
predatory behaviour *see* hunting
pregnancy 120–123, 136
    changes during 63
    choosing nest site 62, 122–123, **123**
    diet/nutrition during 121
    gestation period *61*, 62
    health checks during 121, **121**
    and parturition *see* parturition
    and pre-natal stress 67, 85, 120–122
    pseudo- 60, 62
prenatal stress 67, 85, 120–122, **121**
Prionailurus bengalensis 4, *10*
prospective owners, advice for 129–130, 131–139
    for adult cats/older kittens 138
    bringing cat/kitten home 138–139
    and choice of kitten/adult cat 133–134, **135**
    and choice of male/female/neutered/entire 135
    and choice of pedigree/non-pedigree 135
    cost considerations 132
    and early-life influences 136
    and house-training *see* house-training
    and households with dogs/children 137, **137**
    for kittens 136–138
    and new introductions 133
    and owners' responsibilities 134
    space requirements 131–132
proteins 51–52, 83, *175*
pruritus 81
pseudopregnancy 60, 62
psychogenic alopecia 86
punishment 109, 110, 150, 201, 245, **246**

pupils
    dilated 46, 93, 153, 158, 256
    elliptical 14, **15**
purring 22, 24, 35
    by kittens/nursing females 26
    as sign of immanent parturition 123
    solicitation/unsolicitation 24
puzzle feeders/foraging games 125, 182, 195–199

queens
    age of puberty of 60
    aggression of 59, 65
    and breeding *see* breeding
    in feral colonies 43
    mating behaviour 62
    nutritional status of 67
    oestrus cycle of 60, *61–62*
    pre-weaning period 64–65, **64**, 123–124
    pregnant *see* pregnancy
    in social groupings 43
    and spraying 38
    vocalizations of 22, 25, 26
    and weaning *see* weaning
    see also mother–kitten relationship

RCVS (Royal College of Veterinary Surgeons) 168
reflex reactions 100
reinforcement schedules 105
religion and cats 7–8
renal failure 81
referral 167–168
    and RCVS code of conduct 168
rescue shelters 86, 133, 180–187
resting places 170, 190
    in multi-cat households 45, 47, 206, 214
retina 14, 16, *175*
righting reflex 19–20, **20**
rodent/vermin control 6, 7, 10, 42
rods/cones 14, 15–16
Royal College of Veterinary Surgeons (RCVS) 168
rubbing behaviour 59, *61*
    see also allorubbing
rusty-spotted cat (*Prionailurus rubiginosus*) 4

saccades 15
saccular otolith organ 19
'safe room' for new cats 138–139, 211–212, 219
Saguinus bicolor 25
salivation 93, 256
SAM (sympathetic–adrenal medullary) axis 83

sand cat (*Felix margarita*) 4
Scottish Fold **9**
Scottish wildcat (*Felix silvestris silvestris*) 3, 5
scratch posts/pads 37, 112, 113, 182, 184
scratching 36–37, 112, 250
    and medication 173, 174
secondary/conditioned reinforcement 103–106, 107
    basic principles of 105–106
    and cat carriers 236–237
    and luring **106**, 106, 108
    and reinforcement schedules 105
    and rewards 104
    and sound/clickers 105–106, 107, 113
semiochemicals 35, 174
senses 14–20
    balance 19–20, **20**
    hearing *see* hearing
    sight *see* vision
    smell *see* olfaction
    touch 18–19
sensitive period 69, 70, 72, 126
sensitization 100
separation anxiety 50, 51
serotonin 170, *171*, 176, 177
serval (*Leptailurus serval*) 4, 10
sexual behaviour 58–62
    aggression/fighting 58
    of feral cats 43
    mating 59, 62
    and pheromones 36
    of queens *see* queens
    spraying 38, 58
    of tomcats 58–59
    vocalizations 22, 25, 58, 59
sexual marking 59
shaping behaviour 107–108
    and cues 108
shelters *see* catteries/shelters
shelving/elevated areas 74, 120, 132, 149, 154, 157, 181, 193, 195, 212, **212**, **219**
short-haired breeds 8
Siamese, cross-eyed 16
sight *see* vision
'sit' command **106**, 108, 114
skin diseases 86, 87
skin glands 36–37, **37**
sleeping, feigned/defensive 94, 158, 256
social behaviour 42–51
    bonding *see* social bonding
    effects of neutering on 44
    of feral cats 42–44
    in multi-cat households *see* multi-cat households
    and neighbourhood cats 49
    with people 49–51, **51**, 71

social bonding 45–47, **46**, 49
    misinterpretation of signs of 46–47, 251–252
    signs of 45–46
social play 72–73, 73, **73**, 75
social roll 31, **32**, 254
socialization 70–71, 187
    and handling 49, 71, 127, 128, **128**
    with other animals 70, 127
    with other cats 44, 70, 126–127
    with people 71, 127–128, 136, 150
    and pheromonatherapy 173
    and relatedness 70
    of wild cats 3
spaying 44
spitting 25, 46
stress 11, 82–94, 120, 141, 145
    acute 92
    assessing 92–93
    and cardiovascular system 90, 153
    and cat cafés 191–192
    and catteries/shelters 86, 182, 183, 184
    and compulsive behaviours 92
    and dermatological conditions 86–87
    and electric containment systems 142–144
    and endocrine system 88–89
    and enforced proximity/resource
        sharing 42
    and eustress/distress 82
    and FIC 84–85
    and gastrointestinal system 87–88, 91, 92
    in hospitalized cats 158
    and HPA/SAM axes 83
    and immune system 85–86
    and indoor cats 141, 142, 147, **147**, 148
    and lifespan/ageing 90
    medication/therapies for 170–173, *175*,
        176–177
    and mental/emotional health 90–92
    and multi-cat households 45
    and neurological system 89–90
    diagnostic parameters influenced by 153
    observable signs of 93–94
    of owners 151, 152, 155
    and physical health 84–90
    physiological response to 82–84
    prenatal 67, 85, 120–122, **121**
    signs of, physiological/behavioural
        256
    and spraying 38–39, 85, 91
    and visits to vet's 151, 152, 153, 162
    see also anxiety
stroking 35, 50, 71, 157, 253–254
    benefits to humans of 11
    and training 104, 112
successive approximation *see* shaping behaviour
supercilliary whiskers 18
superstitious behaviour 114

tactile communication 33–35
    allogrooming 33, 34–35, 45
    allorubbing 33–34, **34**, 45
    sniffing/nose touching 33, **33**
    tail wrap 34
tail, signalling with 29, **30**, 93, 256
    'tail up' 29, **30**, 33
tail wrap 34
talking to cats 50, 127, 157, 253
tapetum lucidum/cellulosum 14
taste buds 52
teaching to come to owner 226
teeth 54, 89, 243
territory marking 37, 39, 91
TICA (The International Cat Association) 8
tiger (*Panthera tigris*) 3, *4*
toileting
    of kittens 69
    and pain 80
    see also litter trays
tomcats 58–59
tongue, signalling with 28, **29**
touch, sense of 18–19
toys 106, 107, 113, 120, 201–202, **248**
    bat/chase 201–202
    lights/laser pens/shadows 202
    motorized 201
    wand 50, 104, 193, 202
training rewards/reinforcers 102–103, 104
travel and stress 100, 101, 111, 151, 152,
    153, 185
    see also cat carriers
TTouch 177

unconditioned response/stimulus (UR/US)
    100–101
United States (USA) 173
    cat breed registries in 8
    cat ownership in 10
unwanted behaviour, modifying 109–111
    and extinction/extinction burst 110, 111
    and motivation 112–113
    and non-reward 109–111
    and punishment *see* punishment
upper respiratory infection 85–86
UR/US (unconditioned response/stimulus)
    100–101
urinary diseases 81, 84–85, 86, 91
urination
    indoor 245–247
    and pain 80
    and stress 93
urine spraying 37–39, **38**, 120, 145, 256
    hypotheses for 39
    and neutered cats 91
    as sexual marking 38, 58, *61*

urine spraying, indoor 38–39, 85, 91, 211, 234
    medication for *172*, 173, 178
    first aid advice for 245–247
utricular otolith organ 19

vaccinations 120, 128, 129, 138, 151
vacuum cleaners 71–72
vermin control 6, 7, 10, 42
vestibular system 19
veterinary care 132, 151, 152
    and owners' reluctance 151, 152
    and sensitization 100
    teaching cats to tolerate 243–244
    see also cat-carriers
veterinary professionals, advice for 152–178
    and appointments system 154
    and behavioural consultation *see* behavioural consultation
    and client information 154
    and client interaction *see* veterinary professionals–client interaction
    and clients' stress 152, 155
    and consultation room 155
    and handling/examining 155–157, **156**
    and hospitalization *see* hospitalized cats
    online 164
    removing cat from carrier 156–157, **157**
    and shelving/elevated spaces 154, 157
    and waiting room/reception area 153–155
veterinary professionals–client interaction 164–167
    and behavioural first aid *see* behavioural first aid
    handouts 165
    and nurse/technician behavioural clinics 164
    public talks 164
    staff greetings/reception 154–155
vibrissae 18
vision 14–16, **15**
    binocular/3D 16, 54
    colour perception 15–16
    field of 16
    focusing 15
    in indoor-raised cats 69
    in kittens 68–69
    movement detection 15
    night/low light 14–15, 54
    short-/long-sightedness 69
visual signalling 22, 26–32
    advantages/disadvantages of 23

cats/dogs compared 26–27
    with ears 27, **28**, 93
    with eyes 27–28, **28**
    interpreting, precautions with 27
    with tail 29, **30**, 93
    with tongue 28, **29**
    with whiskers 28–29
    with whole body *see* whole body signals
vocalizations 22–26, 68, 158
    advantages/disadvantages of 23
    chatter/chitter 22, 25
    defensive/antagonistic sounds 23, 25
    hissing 26, 46, 201, 213, 214, 254
    of kittens/nursing females 26
    meow *see* meowing
    murmur sounds 22
    purr *see* purring
    sexual 22, 25, 59
    strained intensity sounds 23
    vowel sounds 22
vomeronasal/Jacobson's organ 17, **18**, 35, 68
vomiting 88

walkways 193, 195, **197**
wand toys 50, 104, 193, 202
watching behaviour 47
water dishes 121, 184, 206, 214
weaning 52, 65–66, **65**, 121, 124
    early, and pica/wool-sucking 92
    frustration during 66, 125
    and hand-rearing 124–125, **125**
whiskers 18, 28–29
whole body signals 30–32, **31**, 93, 247, **247**, 256
    distance decreasing 31, **32**
    distance increasing 30, **31**
    resting positions 32, 46, **46**
wild cats *see* feral cats
    wildcat (*Felix silvestris*) 3, 4, 5
    see also African wildcat
wildcats (*Silvestris* spp.) 5
wildlife 54–55
witches/witchcraft 8
wool-sucking 92

yowl/caterwaul 22, 23, 25, 58

zoopharmacognosy 176, 177

CABI – who we are and what we do

This book is published by **CABI**, an international not-for-profit organisation that improves people's lives worldwide by providing information and applying scientific expertise to solve problems in agriculture and the environment.

CABI is also a global publisher producing key scientific publications, including world renowned databases, as well as compendia, books, ebooks and full text electronic resources. We publish content in a wide range of subject areas including: agriculture and crop science / animal and veterinary sciences / ecology and conservation / environmental science / horticulture and plant sciences / human health, food science and nutrition / international development / leisure and tourism.

The profits from CABI's publishing activities enable us to work with farming communities around the world, supporting them as they battle with poor soil, invasive species and pests and diseases, to improve their livelihoods and help provide food for an ever growing population.

CABI is an international intergovernmental organisation, and we gratefully acknowledge the core financial support from our member countries (and lead agencies) including:

UKaid from the British people | Ministry of Agriculture People's Republic of China | Australian Government Australian Centre for International Agricultural Research | Agriculture and Agri-Food Canada | Ministry of Foreign Affairs of the Netherlands | Schweizerische Eidgenossenschaft Confédération suisse Confederazione Svizzera Confederaziun svizra — Swiss Agency for Development and Cooperation SDC

Discover more

To read more about CABI's work, please visit: **www.cabi.org**

Browse our books at: **www.cabi.org/bookshop**,
or explore our online products at: **www.cabi.org/publishing-products**

Interested in writing for CABI? Find our author guidelines here:
www.cabi.org/publishing-products/information-for-authors/